Powder Metallurgy
The Process and its Products

New Manufacturing Processes and Materials

Powder Metallurgy

The Process and its Products

Gordon Dowson

Adam Hilger, Bristol and New York

British Library Cataloguing in Publication Data

Dowson, Gordon
Powder metallurgy.
1. Powder metallurgy
I. Title II. Series
6713'7

ISBN 0-85274-006-9

Library of Congress Cataloging-in-Publication Data

Dowson, Gordon.
Powder metallurgy: the process and its products/Gordon Dowson.
p. cm.—(New manufacturing processes and materials)
Includes bibliographical references.
ISBN 0-85274-006-9
1. Powder metallurgy. I. Title. II. Series.
TN695.D69 1990
671.3'7—dc20 89-24564
CIP

Series Editor: **John Wood**, Nottingham University

Published under the Adam Hilger imprint by IOP Publishing Ltd
Techno House, Redcliffe Way, Bristol BS1 6NX, England
335 East 45th Street, New York, NY 10017-3483, USA

Typeset by BP Integraphics Ltd, Bath, Avon
Printed in Great Britain by J W Arrowsmith Ltd, Bristol

Contents

Series Editor's Foreword

It is becoming widely recognized that courses in Materials Science and Engineering at colleges, polytechnics and universities must be seen in the general context of manufacturing. Conversely the lip-service often paid by engineering students to materials is now seen as an Achilles' heel in the training of personnel who have to make design decisions. There is still a necessity for research and teaching in the fundamental physics and chemistry of materials (referred to as materials science and covered in physics, chemistry and biological sciences, in addition to metallurgy and materials science departments) but there is a genuine gap when considering materials design and engineering. The aim of this series is to provide state-of-the-art books which address the interface between materials manufacturing and design. They are not intended to be treatises on the fundamentals of materials, but rather designed for students who appreciate the need to quantify materials performance in terms of fundamental parameters but need to use them in, and for, real situations and applications.

Educationalists in universities are extremely concerned that graduates in manufacturing-related courses should be made aware of the implications of the material, the manufacturing process, and the design criteria on the final object. It has to be said that this area is covered extremely well in a number of industrial countries by the use of a team approach to the design process. In the UK, recent surveys have demonstrated that there is an appalling lack of knowledge about materials and processes among design engineers in many industries. It is this situation that is now blamed for much of the lack of a competitive edge in the application of basic science to real life problems. The authors in this series have been asked to tackle both materials and manufacturing processes in this context. The books are intended to be of interest to students on engineering, and specifically manufacturing engineering, courses. In the main they will be written by people who have considerable experience of the industrial side of engineering and as such will not require high levels of mathematical skills. In addition, they are intended to be useful as way-in texts for practising engineers who need to make themselves acquainted with a new field or new material in a context which is relevant to them. The success of the series will be measured by the extent to which design engineers and production engineers are influenced to make new products in an innovative way.

John Wood
Nottingham University

Acknowledgments

The author would like to thank the many manufacturers of sintered components who have contributed items for inclusion in the section of Examples and illustrations for the text; and additionally powder manufacturers Höganäs AB, Norddeutsche Affinerie, and SCM Metal Products for pictures and technical information.

I am especially pleased to acknowledge the very great help and encouragement that I have received from the staff of *Metal Powder Report*, particularly Mr B Williams.

A G Dowson

Introduction

Powder metallurgy, commonly designated by its initial letters as PM or P/M, may be defined as the production of useful artefacts from metal powder without passing through the molten state. It should include at least some consideration of the processes by which powders are produced, because the behaviour of the powder in the subsequent consolidation stages can be markedly dependent on the properties of the powder which, in turn, are influenced by the production process employed.

Thus we can distinguish two important steps in PM:

(a) the production of the powder, and
(b) its consolidation.

Both these topics are treated in subsequent chapters, but it will be useful at this stage to give an outline of the sequence of operations following the manufacture of the powder in suitable form.

The basic steps are threefold.

(1) Mix or blend the powders, normally incorporating an organic substance to act as lubricant during the next step, and test to ensure conformity with predetermined standards.

(2) Load the powder mix into a suitable die or mould and consolidate it by the application of pressure into what is referred to as a *compact*, analogous to an aspirin tablet. Compacts must have sufficient strength to permit handling without fracture or crumbling, but are in no way strong enough for any engineering application.

(3) Heat the compacts, generally in a protective atmosphere, to cause the particles to weld together thus generating the strength required for use. This heating process is called *sintering*, and components made individually by PM are, therefore, referred to as

sintered parts or *sintered components*. Another term in common use is PM *parts*, and these three terms are used indiscriminately in the literature of powder metallurgy.

The sintering temperature is normally below, and in many cases significantly below, the melting point of the metal, and the welding together of the particles occurs by atomic diffusion; hence the term *solid state sintering*. However, in some important cases where the metal concerned is an alloy having a melting range, as against a fixed melting point, sintering may be carried out at a temperature sufficient to generate a small amount of liquid, which, as could be expected, markedly assists the diffusion process. The process is then referred to as *liquid phase sintering*. Since the essence of the PM process is, in most cases, the production of articles having substantially the required finished shape, it follows that only small quantities of liquid phase can be tolerated, otherwise the object would be in danger of warping or otherwise losing its shape. A typical example of liquid phase sintering is provided by the widely used mixture of iron powder with a few per cent of copper powder, such mixtures being normally sintered at about 1120 °C, i.e. some 40° above the melting point of copper.

In certain special cases steps (2) and (3) are combined, that is the powder mix is subjected simultaneously to pressure and elevated temperature; this process being referred to as *hot pressing* or *pressure sintering*.

In many cases the component is, after sintering, subjected to additional processes: re-pressing to correct the dimensions (an operation that also increases the strength), surface treatments such as case hardening or plating, barrelling to remove burrs etc. These *post-sintering processes* are the subject of a separate chapter. The various stages of the PM process, including some not already mentioned, are shown diagramatically in figure 1.

The Historical Development of PM

By stretching a point the origins of powder metallurgy can be traced back to antiquity when man's ability to melt the metals then available—copper and iron, silver and gold—was limited. Instead of

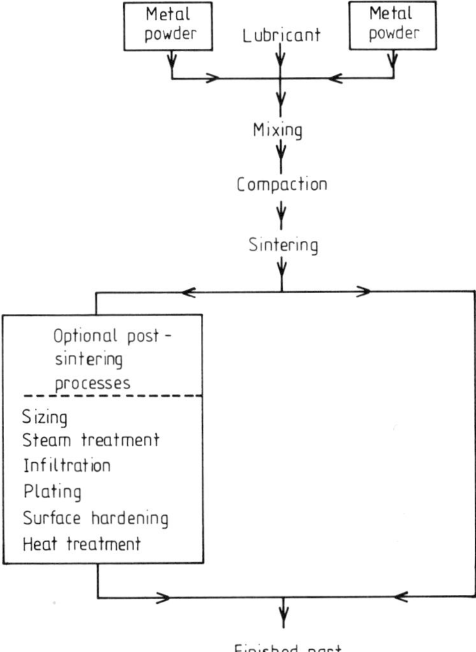

Figure 1 Sequence of operations in the production of sintered components.

melting, the craftsmen consolidated naturally occurring nuggets of the precious metals and sponges of metals that could readily be produced by the reduction of their ores at low temperatures, by hammering at either ambient or elevated temperature. This process, with many modifications, was used for iron well into the present century, and many fine examples of wrought iron are still extant. The term has lately been somewhat debased to refer to supposedly decorative articles made from wrought mild steel.

The next flowering of PM followed the introduction of platinum brought by the conquistadores from South America. This metal could not be melted, but in the early part of the 19th century workers in England, Spain, and Russia developed similar processes for making wrought platinum—largely for use as crucibles in the chemical laboratory—by what was fundamentally a modern PM route. Wollaston has been referred to as the father of powder metallurgy.

The next significant development was with another metal of even higher melting point—tungsten. This metal was, and still is, of interest for use as the filament of incandescent electric light bulbs which up until then had been made of carbon, which, of course, is extremely brittle. Unlike the earlier quoted examples of PM products, wrought tungsten is still made from powder. It is referred to more extensively in the next chapter.

A second important product that is still very much with us was, like tungsten wire filaments, pioneered in the USA, and is porous bearings. Unlike the earlier products that were made from powder because the metal concerned could not readily or at all be processed by melting, porous bearings are made by PM because of the special properties that result. The bulk of these components are of bronze (10% copper, 10% tin), which can be melted quite easily; and indeed, cast bronze is in general use as a bearing material. When the bearings are made by PM, however, it can be arranged that a considerable volume of interconnected porosity remains, and, if the gas is extracted from the pores and the parts immersed in a lubricating oil, the pores are filled with oil. Such parts are used as bearings in a wide range of small rotating or reciprocating machinery and, at best, require no further lubrication during the life of the equipment. When did you last oil your vacuum cleaner? Such oil-impregnated parts are variously referred to as *self-lubricating* or, by purists, as *oil retaining bearings*. The volume of retained oil is commonly between 20 and 25% of the total volume.

At about the same time, i.e. shortly after the First World War, another important PM product made its debut. In 1925 Shröter, of the German company F Krupp, was granted a patent for a process and product consisting of tungsten carbide particles held together by a 'cement' consisting of metallic cobalt. This material was used originally in the form of wire drawing dies, for tungsten, as a replacement for diamond dies. Hence the trade name 'Widia' from the German Wie Diamant (like diamond). It was subsequently found that this *cemented carbide* or, as it is now called, *hardmetal*, was extremely useful also as a cutting tool material, and engineers are all familiar with the term 'carbide tools'. Binder phases other than cobalt, which is expensive and comes mainly from what are called politically unstable regions, have been investigated very extensively but have, so far, found only limited application. However, many alternative and additional hard particles have been found to give improved properties. These include the carbides of other metals—titanium, tantalum, niobium, etc—as well as nitrides and borides. Detailed discussion of cemented carbides and the like is beyond the scope of this book, but the shortness of this paragraph should not be allowed to obscure the fact that the hardmetal industry now has a turnover very much larger than the whole of the rest of the PM industry.

Sintered Components

It was only after the Second World War that the products with which we are here mainly concerned, those conjured up by the term sintered parts, were developed on a commercial scale. Unlike the PM products touched on above, for which powder metallurgy is for one reason or another the only practical manufacturing route, the post-1940 development was of components that could be made more or less readily by traditional processes—casting, forging, stamping, or machining from bar stock. The justification for making them from powder was almost exclusively a question of cost. Reference to the examples of present day components illustrated at the end of the book shows that even now in the majority of cases cost saving is the principal or only advantage claimed for making the part from powder. However, in a number of cases improved performance of the component is claimed, for instance example 31, a mould for glass bottles, and example 35, a high speed steel threading die, where PM gives a better as well as a cheaper product. The important question of cost is explored in some detail in the next chapter. However, the product that could be said to have triggered off the development had its raison d'être elsewhere. Western Europe has very limited deposits of copper ore and the shortage of copper for driving bands for artillery shells led to a search for possible substitutes. Solid iron or steel is unsuitable by reason of its hardness, but research proved that sintered—and therefore porous—iron performed satisfactorily.

In spite of this success, PM's first steps were very hesitant: at a meeting in London one expert went on record as doubting whether sintered parts could ever be of use in load bearing applications. Subsequent developments have shown how wide off the mark such a judgement was. During the next four decades improvements to the quality and compositions of available powders, the development of better compacting presses, improved compacting tool design, better sintering furnaces and furnace atmospheres, together with a wide range of post-sintering treatments have led to a general acceptance of sintered parts not merely as a low cost substitute for parts made by other processes, but in several instances as providing better properties in the finished product than can be achieved by traditional means. These developments are the subject of a later chapter.

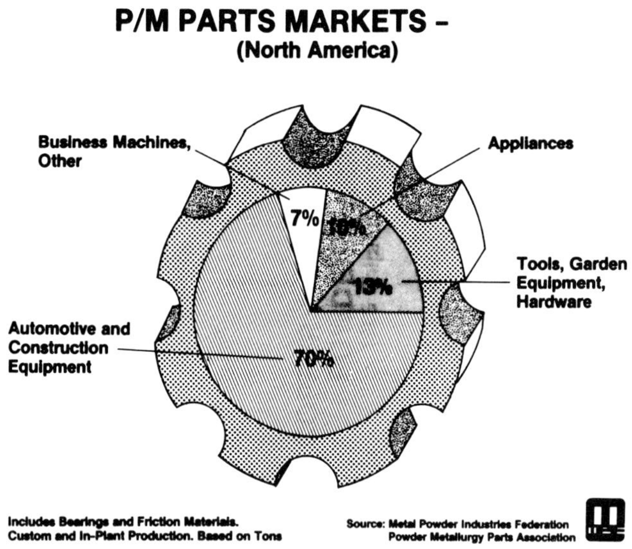

Figure 2 The principal users of sintered components.

It will be interesting at this stage to take a look at the sort of parts that we are talking about, and where they find application. Taking the latter question first, a fairly accurate guide to the market is given in figure 2 from which it is clear that motor cars are by far the biggest single consumers of sintered parts. Sustained efforts have been and are being made to develop other markets and lessen the dependence of the PM industry on the car industry, and these are having considerable success, but currently the prosperity of the PM industry is inescapably bound up with that of the automotive industry. Figure 3 shows a very small selection of parts currently in service, and Figure 4 some of the parts used in two other applications, and these provide an answer to the first question—what sort of parts are made by PM? To attempt to list them all would be an impossible task; there are manufacturers with a stock of pressing tools numbering as many as 10 000, although they are not all in use at any one time; and one company is introducing new parts at the rate of more than one hundred in a single year.

Figure 3 Some parts used in motor cars. *Mannesmann*.

Figure 4 High strength parts used in textile machinery and woodworking equipment. *Metalsinter SRL.*

For reasons that will emerge in a later chapter, PM parts are normally quite small. In one company, admittedly an exceptional case, the average weight of parts made is of the order of 10 g, but though this is exceptional, the vast majority of sintered parts weigh less than 1 kg. The examples show only one part weighing more than 1 kg and the part concerned, example 34, is a special case, namely a zirconium valve body, not a run of the mill structural part, and note also that iso-static compaction which largely removes the size limitation was used (see chapter 4). However, parts of increasing size are gradually appearing—the world's largest to date weighs about 4 kg—but, as will be explained later, there are severe technical as well as economic restraints on the size of part that can be produced directly by pressing and sintering. As regards shape, the examples show that the diversity is enormous; some parts have what appears to be very simple geometry while at the other end of the scale extremely complex shapes are possible. The PM process is especially suited to the economic production of complicated shapes because not least among its merits is the ability to produce the required shape and dimensions without recourse to machining. The elimination or reduction in the amount of machining is quoted in many of the examples, e.g. examples 4, 22 and 29. Toothed components are particularly relevant in this context.

There are limitations to the shapes that can be produced, relating principally to the requirement that the compact has to be ejected in one piece from the die, but so long as this is taken into account at the design stage it is not an unsurmountable handicap. This aspect and some of the ruses that are available for overcoming the limitations are covered in later chapters.

The above refers mainly to what are termed mechanical parts which, by any reckoning, is by far the biggest group. The bulk consists of iron or steel, which we lump together as ferrous parts, but significant tonnages of parts in copper, brass, bronze, and aluminium are produced, and sintered titanium alloy parts have recently made their appearance. However, in addition to mechanical parts others produced on a large scale include self-lubricating bearings, magnetic components, electrical contacts, filters, friction materials for clutches and brakes, and, as indicated earlier, hardmetal parts for metal working tools, wear parts generally, and especially for metal cutting tools.

Related to hardmetals is a good example of another category of PM products that is becoming of increasing importance, namely *cermets*, by which is meant a composite material having ceramic particles in a metallic matrix. Hardmetal is, in fact, in this category, but is a very special case.

A further and quite separate area of PM that is developing rapidly is the production of wrought material from billets or the like. These powder-based billets are then further processed in the same way and using the same machinery as ingot-based billets, i.e. extrusion, rolling, press-forging etc. The justification for starting from powder lies principally in the improved or special properties that can be achieved in the finished product. The size restriction applicable to directly pressed and sintered parts does not apply in these cases for reasons that are explained in chapter 10.

1 Why Make Things From Powder?

As has been indicated in the introduction, there are several, largely unrelated, reasons for making engineering components from powder. Leaving aside the cemented carbides which constitute a class apart, the biggest sector of PM, whether the yardstick be tonnage or value, is the direct manufacture to finished size and shape of structural components for machinery; and in this sector the justification for adopting the powder route is, with a few exceptions, economic, that is at the end of the day the PM part is cheaper. It is appropriate, therefore, to explore this aspect of the subject first.

The Economic Case

Except for a few special cases, powder is not the normal product of the process of extraction of a metal from its ore. In the case of iron which constitutes the bulk of PM structural parts, the usual process is via the blast furnace which produces (very impure) liquid iron—pig iron—that is then refined by a steel making process and cast into ingots. To convert this metal into powder involves an additional and very significant cost. Furthermore, the starting material for the PM process cannot be just any sort of iron powder; composition and physical characteristics must be closely specified and adhered to. These factors are explored in some detail in chapter 3; the purpose of mentioning them here is to indicate why, inevitably, iron powder in a suitable form is significantly more expensive than the raw materials that are the starting point of alternative shaping processes for engineering components. It may be mentioned that one widely used process for the production of iron powder does not go via the blast furnace/steel refining route. Where an exceptionally pure grade of iron ore is available, it can be directly reduced to sponge which is subsequently crushed to powder. However, as will be seen later, the process is not quite so simple as it sounds, and a number of intermediate steps are required in order to meet the rigid specifications laid down by the PM industry. The availability of this process does not at all invalidate the statement made above that iron powder suitable for PM is very much more expensive than other forms of ferrous raw materials.

How then is it possible, with this handicap, to produce the final component at a lower cost than that for competing processes? The answer is twofold. Firstly, much better material usage, i.e. the weight of the finished component is a much higher percentage of the weight of the starting material consumed, sometimes by a factor of two or more. Secondly, and largely responsible for the first, the PM process can produce the finished part directly to the required dimensions without any subsequent machining or at worst with very much less machining. This not only contributes largely to the better material usage by eliminating or markedly reducing the amount of scrap, but also saves the cost of machining which can be very high for complicated shapes. Recent careful calculations have shown also that the total energy consumed in all the stages of manufacture of ferrous components from the ore to the finished part is least for the PM route, and since it seems unlikely that the relative cost of energy will do other than increase with the passage of time, this advantage of the powder process will remain and even increase in importance.

The above discussion has referred mainly to ferrous materials, but broadly speaking, the same factors apply to non-ferrous metals. Two differences, however, are worth mentioning. The common non-ferrous metals, copper and its alloys bronze, brass, and nickel-silver, and also aluminium, are relatively very easy to fabricate from solid, thus reducing the advantage of PM. Against this it may be noted that these metals are very much more expensive than iron and, therefore, the raw material utilization factor is of increased importance. This applies, for example, to the zirconium valve body, example 34, and to the titanium compressor link, example 33. Nevertheless, the volume of non-ferrous structural components is but a small fraction of the total for ferrous materials.

Consideration of the foregoing will bring out two important features: the economic advantage of the powder route will be greater

(a) with smaller components where the raw material cost is a smaller proportion of the total cost, and

(b) with more complicated shapes that require a large amount of machining and expensive machine tools when they are made from ingot-based metal.

Apropos of size, there are other factors that put a brake on the continuing trend to bigger and bigger parts. Firstly, bigger parts require bigger, more powerful, and, therefore, more expensive processing equipment—compacting presses and to a lesser extent sintering furnaces. Secondly, as will be evident when we come to look at the compacting and especially at the sintering processes, certain technical difficulties increase in importance as the size of part is increased.

The Special Cases

While acknowledging that economic considerations predominate for the bulk of PM parts, there are several groups of materials of great importance to engineers where the justification for using PM is to be sought in technical aspects either of the manufacturing process or of the properties of the product.

Group 1. Refractory metals

It was mentioned in the introduction that one of the earliest PM products was tungsten wire. The particular characteristics of tungsten that dictated the use of powder metallurgy were:

(a) the extraction process culminating in the reduction by hydrogen of the oxide yields the metal as a powder;

(b) the melting point of tungsten, in excess of 3000 °C is so high that normal melting equipment could not cope with it, and furthermore no refractory material was available for the manufacture of crucibles to withstand such temperatures;

(c) when melted material was eventually produced it was completely brittle, and, therefore, could not be shaped by any known mechanical working process—forging, rolling, wire drawing etc.

Modern melting procedures, arc and electron beam melting on a water-cooled hearth, have made it possible to produce 'ingots' of tungsten, but the brittleness mentioned above still precludes the fabrication of wrought shapes. PM remains, therefore, the accepted commercial process for making tungsten wire and sheet. For electric lamp filaments a dopant, commonly thoria, is added before the reduction stage. Bars of the doped powder some 2 cm square are pressed in special dies, and the compacts which are very fragile are carefully pre-sintered at ~1200 °C in hydrogen to give them sufficient strength to be handled. Final sintering at up to 3000 °C is done by passing electric current through the compact which is clamped in a vertical position in a hydrogen atmosphere. A density of about 90% theoretical is aimed at. The sintered bar is then hot swaged at progressively lower temperatures, during which process the grains assume an orientation favourable to further extension in the longitudinal direction until, finally, the wire can be drawn at temperatures from 800 °C down to about 400 °C. It is postulated that the mechanical working is made possible by the fact that the bar is initially porous and, therefore, initial deformation is partly absorbed in closing the porosity and partly in inducing the preferred orientation of the tungsten grains. The function of the dopant referred to earlier is to assist in the retention during service of the very markedly elongated microstructure induced by the swaging and wire drawing, which structure is less brittle than the equiaxed grain structure which would otherwise develop in a light bulb filament. The principle has more recently been applied to other metals for use at high temperature, e.g. platinum.

Other refractory, i.e. high melting point, metals—molybdenum, tantalum, niobium, rhenium, and the platinum group metals osmium, ruthenium and iridium—were in the same category as were the somewhat lower melting point metals titanium and zirconium. These last two and tantalum are now regularly made by fusion metallurgy, but for the rest PM is the preferred route. Indeed, certain tantalum products, foil for example, are still made from powder, and there is great interest in PM-produced titanium alloys which are referred to in a later chapter.

Group 2. Composite materials

In the present context there are two types of composite. Firstly, mixtures of two metals that are substantially insoluble in each other and therefore cannot be produced by melting and casting. Secondly, and of increasing importance, mixtures of a metal with a non-metallic substance such as an oxide or carbide of a metal, often as a fine powder. The second phase may alternatively be in the form of fibres. In this group should be included diamond cutting tools and cemented carbides, but as these have been dealt with in the introduction they will not be considered further.

Among the PM composites of industrial importance are the following.

(a) *Electrical contact materials*. These are for the most part based on a metal with very high electrical conductivity, usually silver or copper. These metals have unimpeachable properties as conductors, but because they are soft, have fairly low melting points, and do not form very stable oxides, they tend to weld together quite readily during use as make and break contacts. If, however, they are mixed with a substance having good arcing and welding resistance, a good approximation to the best of both worlds is achieved. Among the substances used are certain metal oxides, especially cadmium and tin. Cadmium oxide is probably the best from the performance point of view, but it is falling into disfavour because of the potential health hazard. The patent literature abounds with references to alternative oxides and mixtures of oxides. Metallic nickel (with silver only), tungsten or molybdenum and their carbides are other arc resisting ingredients in common use. All these combinations can be made by mixing the ingredients in powder form, pressing and sintering; and since the weld resisting substance is virtually insoluble in the conducting metal, the conductivity of the latter remains unaffected. The overall conductivity of the composite is, of course, somewhat lower than that of the pure metal. With tungsten and molybdenum, an alternative and generally favoured procedure is to press and sinter a very porous skeleton of the refractory metal, the porosity being later filled with molten copper or silver—a process known as infiltration.

(b) *Copper–carbon brushes*. Another very important class of composite sintered contact materials consists of mixtures of copper and graphite in various proportions depending on the intended use. Copper contents of up to 80% or so are used, while at the other end, 70% of the total may be graphite. Other ingredients such as tin and lead may be present in small amounts to modify the properties. All the compositions are made by cold pressing and sintering mixtures of the ingredients in powder form. Those with a high metal content are straightforward PM products, but it is arguable that if 60 or 70% by weight of the material is graphite it should not be classed as a PM material. In support of that view it may be mentioned that a somewhat different process involving the use of organic binders is used in their manufacture.

These metal/graphite compositions are used in applications involving sliding contact, for example in electric motors and generators to which the term carbon brush strictly applies (see figure 5). They are used also to pick up the current from overhead wires to power electric trains.

(c) *Friction materials*. These constitute yet another important class of PM composites. They are heavy duty brake linings and

Figure 5 A selection of carbon brush assemblies. *Norddeutsche Affinerie*.

clutch facings which consist of a metal matrix—copper, bronze, or brass as a rule—in which are embedded particles of a ceramic-type material, often silica. The latter, of course, provides the friction while the metal acts as a heat sink. Additions such as iron, lead, and graphite are often included to modify, i.e. to improve, the performance, for example to make the action smoother.

As these materials are commonly used in the form of sheet having a large surface area, they are very fragile in the as-pressed state, and the manufacturing process is somewhat different from that used for structural components. Additionally, perhaps fortunately, dimensional aspects are much less important.

Group 3. Porous materials

All routine PM materials are porous to a greater or lesser extent, but here we are referring to materials that are required to be porous. There are two important categories: filter elements and oil-impregnated bearings. These products are the subject of a separate chapter.

Group 4. Steel-backed bearings

These constitute another important class of bearing materials made by PM. They are used extensively for main and connecting rod bearings in motor cars etc, and consist of a relatively thin layer of bearing metal sintered onto a steel strip which is finally cut and shaped into half-bearings and fitted in suitable housings. Many different compositions of bearing metal have been used including Babbit, aluminium/lead alloys, copper/lead and bronze. These lead bearing alloys provide, incidentally, good examples of composites of two metals that are insoluble in each other in the solid state, although they are mutually soluble in the liquid state. So, if the molten alloy is atomized, the resulting powder particles consist of a matrix of copper or aluminium in which is distributed fine particles of lead. Since the lead is insoluble in the solid matrix metal, the duplex structure is retained after sintering.

The process for the manufacture of bearing strip is to spread a uniform layer of powder onto a steel strip, which is then passed through a furnace where sintering takes place and the powder is bonded to the steel. From the sintering furnace the strip continues directly into a rolling mill where the sintered powder layer is densified. It may be made completely dense or left with some porosity to give it oil retaining characteristics. Alternatively, in the case of bronze, the pores are impregnated with PTFE either alone or with lead powder, for use as a dry bearing. The steel strip used is generally copper plated if a copper-based bearing layer is to be applied, to improve the bond between them. The bearing metal powder is normally spheroidal in shape in order to provide dense packing. One manufacturer specifies that the particles shall be potato shaped, explaining that if they are too nearly spherical they run all over the place.

Group 5. High duty materials with improved properties

Finally, and of rapidly increasing importance especially in aerospace applications, are powder-based materials often in wrought form which, by reason of their composition, microstructure, or both, have mechanical properties superior to those of competing materials made from ingot. A separate chapter is devoted to such materials.

2 Powder Manufacture

It is appropriate at this point to ask the question 'What do we mean by Powder?' A first response might well be 'Surely, everybody knows what a powder is', but the question is not quite so simple. Is a collection of 5 mm diameter ball bearings a powder? Surely not, so where do we draw the line? An internationally agreed definition arbitrarily specifies a maximum dimension of 1 mm (ISO 3252).

There are very many ways in which metals can be produced in powder form, and a very large book could be written on the subject, even if it included only those processes that are of commercial importance at the present time. In fact there is a whole book devoted to a single production process—atomization. We have space here only to review the subject and attempt to highlight the factors that are of particular importance to the subsequent use of the powder in the manufacture of engineering components.

Comminution

The mechanical disintegration of solid metal into small pieces is an obvious powder making process; iron filings provide a good example. Dental amalgam alloys, basically silver/tin alloys, were, until recently, made exclusively in this way, originally by filing and later by controlled turning of a bar on a lathe. Comminution has not been of great importance for the production of the basic powders for the manufacture of sintered parts, but repeated attempts have been made to use swarf either directly or after further comminution to powder. In 1980 Lenel wrote 'the process has not been widely used commercially', but more recently the successful use of this relatively cheap material, both steel and cast iron, on an industrial scale has been reported. The process can be expected to increase in popularity but there is a limitation. In order to get reproducible properties on which the success of the PM process depends, it is essential to have an assured supply of swarf of uniform known composition and uncontaminated by extraneous solid matter. Floor sweeping would be highly suspect.

However, ball milling is the standard process for the production of powders of brittle materials such as ferro-alloys (ferro-chrome, ferro-manganese, ferro-silicon), which are now being used as additions to iron powder mixes. Ball milling is an essential part of the process for the production of fine powders of the carbides of tungsten and other refractory metals and their carbides during the manufacture of hardmetals.

Another comminution process, the Coldstream process, is used especially for the manufacture of very fine powders. Granular material is entrained in a stream of gas at high pressure which is then adiabatically expanded through a venturi nozzle. This cools the gas and at the same time the metal particles which are thereby embrittled. On emerging at high speed from the venturi the particles impinge on a suitable target, which may be of the same material thus avoiding any risk of contamination. Here the brittle particles fracture into smaller fragments. The gas may be air, but an inert gas is used if the metal being processed needs to be protected from oxidation. The required fine powder fraction is separated from the emerging gas stream and the oversize is recycled so that the eventual yield is practically 100%. Commonly, particle sizes of 10 μm or less can be produced. The process was originally used mainly for the recovery of hardmetal (cemented carbide) scrap, but the recently developed injection moulding process which needs very fine powders for its success is likely to extend its use considerably. Coarsely atomized spherical powder can be used as feedstock.

Comminution is extensively used as one step in other powder production processes, that is for breaking up metal sponge or cake resulting from oxide reduction processes.

Solid State Reduction

For many years this has been the most widely used process for the manufacture of iron powder. Pioneered and still extensively used by Höganäs AB in Sweden and by a former subsidiary in the USA, now a jointly owned company, the process depends on the availability in northern Sweden of a reliable supply of very high grade iron ore. The ore is crushed to a size approximating to that of the eventual iron powder, mixed with powdered coke or coal, and passed in refractory tubes through a furnace at a temperature such that reaction occurs, releasing oxides of carbon and leaving more or less pure iron in the form of a sponge which is crushed to powder. If the process were as simple as that, there would not be justification for the price of powder being significantly higher than that of, say, mild steel; in fact, sponge iron produced in substantially the same way is used commercially as feedstock for melting in foundries. For the production of PM-grade powder, additional steps and strict control at all stages are required.

In order to take care of sulphur and other impurities in the coal, an addition of limestone is made to the starting raw material. The reaction is carried out slowly so that the complete furnace cycle takes about 24 hours. During the process shrinkage occurs so that there is no difficulty in extracting the sponge bar from the containers. The sponge is next crushed and the iron particles separated magnetically from the gangue, after which it is passed on a steel band through a second furnace under an atmosphere rich in hydrogen which completes the reduction and anneals, i.e. softens, the particles. However, some sintering takes place and the material emerges once again in the form of a cake which is crushed again to powder and is finally sieved to nominally minus 150 μm. The relevance of the softness of the powder will become clear when compaction is discussed later, but additionally the process must be so controlled that the particle size distribution is consistent and to specification.

Smaller quantities of iron powder are produced by the hydrogen reduction of mill scale. The scale is first crushed to powder and roasted in air to ensure that it is fully oxidized to Fe_2O_3 so that the subsequent reduction will be consistent. As in the Höganäs process, the material is reduced in a band furnace, but pure hydrogen is used. The resulting cake is then crushed and classified. This process is viable only where a reliable supply of mill scale of more or less constant composition can be assured and where a cheap supply of hydrogen is available.

The individual particles of powder produced by the reduction of iron oxide are porous, and such powders are commonly referred to as sponge powders. It will be clear that little or no refining takes place and so the purity of the starting material is a major factor in the success of the process.

There are several other processes in use in which reduction plays a part but where the starting material is mainly an atomized powder. In one process, a carbon-rich iron is atomized, the resulting powder being then mixed with iron oxide in the form, e.g. of mill scale and passed through a band furnace. The carbon and oxygen react to form carbon monoxide, and with proper control, a cake free from both carbon and oxygen is produced. The merit of this and similar processes in comparison with direct atomization, is that the powder particles are once again porous, giving the powder a low bulk density which is advantageous at the compacting stage, as will be seen in the next chapter.

Hydrogen reduction of the oxide is used also in the production of tungsten and molybdenum, and it has been used also for copper. The disappearance of copper scale which has followed the almost universal use of continuous casting put an end to the process, but some powder producers start with atomized copper powder, oxidize it at least partially, and then reduce it. In this way a lower density powder than can be produced directly by atomization is obtained.

Other reduction processes are typified by the Sherritt Gordon process for nickel powder, in which a solution of nickel ammonium sulphate is treated in an autoclave with hydrogen at a pressure of some 1400 kPa. Fine nickel particles are produced and by repeated process steps they grow to the required size. A comparable process has been described for copper.

Copper powder has been, and still is, produced to a small extent by yet another chemical reduction process in which iron is used to throw the copper out of solution, normally of copper sulphate, but chloride also has been used. A recent patent claims the use of scrap aluminium shavings for the same purpose. This process, for some not obvious reason, is known as cementation—hence cement copper. The product, which is in the form of powder, requires further processing before it can be used for PM, and the process is generally confined to the extraction of the metal from low grade residues, mine tailings and the like.

Silver powder may be produced by an analogous process but

is more usually prepared by precipitation from a solution of silver nitrate by the addition of an organic reducing agent or hydrazine.

Electrolysis

This process was at one time standard for the production of copper powder, and was also used for iron before the sponge iron process was developed. In the case of iron, the metal is deposited on stainless steel anodes using cathodes of low carbon iron. The electrolyte could be either sulphate or chloride. The deposit, which is of very pure iron, is quite dense, and by reason of included hydrogen, very brittle so that it can readily be reduced to powder by ball milling. After annealing it is very soft and, as stated, very pure. Nowadays other types of iron powder with adequate softness and sufficient purity are available and so, since the electrolytic process is very much more expensive, electrolytic iron powder is now used only where the highest purity is demanded. Although traditionally the product is referred to as electrolytic iron powder, the pulverization is done by comminution, and it ought perhaps to appear in that section. The same process—pulverization of a dense electro deposit—is used for producing powders of pure manganese and chromium, but these are of only marginal interest in PM.

In the case of copper the electrolytic process was standard until the development of the cheaper atomization process. In this case, however, the process is still very much alive because it can produce grades of powder that are necessary for certain PM products and which cannot be produced by atomization.

In the normal process of copper plating, the objective is to produce a smooth dense deposit on the article being plated, but if the conditions are wrong a rough burnt deposit can result. If powder production is the aim these wrong conditions are what is required. Basically, the conditions for powder production are a low concentration of copper in the electrolyte, which is generally sulphate, and a much higher current density. Typically, a concentration of $50\,g\,l^{-1}$ of copper sulphate and a current density of about $500\,A\,m^{-2}$ are used compared with $150\,g\,l^{-1}$ and $150\,A\,m^{-2}$ for dense copper plating. The electrolyte is continuously circulated through heat exchangers to maintain the temperature at the required figure, around 50 °C. All the parameters mentioned are varied according to the grade of powder required. The soluble anodes are of refined copper and the cathodes on which the copper is deposited are traditionally of lead. The deposit can readily be brushed or shaken from the cathodes, if it has not fallen off of its own accord. Lead was also used to line the vats, but in modern plants these lead-lined wooden vats have been replaced by acid resisting plastic, or rubber-clad steel. Because the anode efficiency is greater than the cathode efficiency, a few dummy anodes of insoluble material are included to avoid a continual increase in the copper concentration.

The copper powder collects at the bottom of the vats and is periodically removed either by means of a slurry pump or by gravity discharge from outlets at the bottom of the vats. It is then filtered and washed free of electrolyte, and dried and simultaneously reduced in a band furnace using an atmosphere of partially burnt hydrocarbon. This process step is necessary because, at the high current densities used, some oxidation of the copper occurs—hence the use earlier of the term 'burnt'. The resulting cake is then ground and sieved.

Apart from the high and increasing cost of electricity, the problem of the treatment and eventual disposal of the large volume of wash water increases the gap between the cost of electrolytic powder and atomized powder, such that some companies prefer to make atomized powder, roast it to oxide, and then reduce it, thus producing low density sponge type powder.

Thermal Decomposition

Certain of the more noble metals form chemical compounds which decompose readily on heating, yielding either powder directly or a sponge that can easily be crushed to powder. Platinum is a good example of the latter, and the normal process for refining this metal yields a pure compound, platinum ammonium chloride, which decomposes to a sponge on heating. Of greater interest to powder metallurgists is the production of nickel powder by the thermal decomposition of nickel carbonyl. This process was developed originally by Mond *et al* about a hundred years ago as a means of purifying nickel. Crude metal is exposed to carbon monoxide gas at elevated temperature and pressure and nickel is selectively converted to the carbonyl—$Ni(CO)_4$—which itself is gaseous at the reaction temperature. By reducing the pressure and increasing the temperature the carbonyl is caused to decompose and deposit nickel in the form of powder or pellets, depending on the process conditions. The metal thus produced is of very high purity, better than 99.99%, and can be extremely fine. It is of considerable importance in the PM industry. The carbon monoxide gas is, incidentally, recycled.

Carbonyl iron is produced by an analogous process and is becoming of some importance as a result of the demand for very fine powders for the injection moulding process, *quod vide*.

Atomization

This somewhat illogical name is applied to any powder process that involves the production of molten droplets by the disintegration of molten metal or by other means, the droplets being allowed or caused to solidify before impinging on a solid surface. A considerable range of processes related only in that they fall within the above definition have been developed, differing markedly in the method by which the molten metal is produced, in the way in which the disintegration is brought about, and how the molten droplets are cooled and frozen. It is well known that drops of a liquid left undisturbed tend to assume spherical shape under the influence of surface tension, and so droplets of molten metal, if allowed to cool slowly out of contact with solid surfaces or with other droplets, freeze as spheres. If on the other hand the droplets are formed by high energy impact of a liquid, they are vastly distorted and, being cooled very rapidly by the liquid, they give solid particles of highly irregular shape. So, by atomization a wide range of particle morphology between the two extremes is possible. We shall see later that, in order to produce compacts of adequate strength, irregular powders are desirable, but for some purposes which are increasing in number and importance, spherical powders are preferred.

A feature of atomization that distinguishes it from alternative processes is that powders of homogeneous alloys can be produced, so-called pre-alloyed powders. This is one of the reasons why atomization is making inroads into the iron powder production field; the sponge powder process gives nominally pure iron. It is worth mentioning in passing, however, that there is a penalty in having pre-alloyed powders: they are inevitably harder than pure metals and therefore less compressible. This important aspect of powders is explored more fully in chapter 4.

Water atomization

This is the most widely used atomization process for the production of PM powders because it yields the highly irregular powders that are required for cold compaction in rigid dies, the process by which the bulk of PM parts is produced.

There are innumerable variations in the detail of the equipment used, but basically, the process is to allow a stream of molten metal to emerge vertically from a tundish into the atomization chamber, where it is disintegrated by the impact of jets of water at high pressure. Pressures range from about 5 to 20 MPa (3000 psi) and, all else being equal, the higher the pressure the finer the powder. Pacific Metals in Japan has recently begun production of very fine powders by water atomization using pressures up to 1000 kg cm^{-2}. Powders of mean particle size 10 μm are being produced. Such powders are ideal for injection moulding, and, of course, fully alloyed powders can be produced. The literature abounds with references to the actual arrangement of the water jets, and many powder producers have their own proprietary designs, but at its simplest and probably most widely used, two flat stream jets located symmetrically on opposite sides of the metal stream are employed, as shown in figure 6. The angle of the jets to the vertical, the volume and pressure of the water, the thickness of the metal stream, and its degree of super-heat, all affect the final result and have to be optimized for each design. Yields of powder of the required fineness—normally minus 150 μm—can, at best, be as high as 95% of the metal melted. Some of the alternatives known to be in use or proposed in the patent literature are four V jets, i.e. two pairs of opposed jets at right angles, a large number of cylindrical jets arranged in a circle, and a circular jet nozzle giving a complete cone of water. A recent patent describes an arrangement of primary atomizing jets surrounded by a curtain of water to increase the cooling rate, the aim being to get still more irregular powder. Regardless of the jet configuration, the general rule is that the greater the volume of water the more irregular the powder.

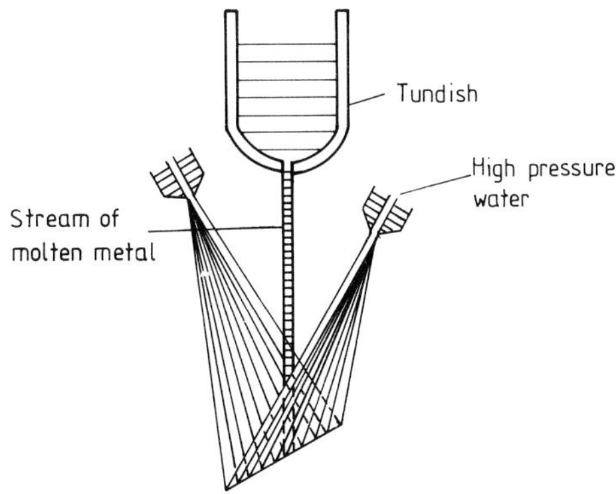

Figure 6 Schematic drawing showing a two-way flat stream jet system for water atomization.

The powder is separated from the water either on a filter of some sort usually involving vacuum, or by settling. This de-watering is followed by drying generally either on trays in forced draught ovens or in a thermo-venturi (flash) dryer. Powders produced in this way inevitably acquire a film of oxide on the surface of the particles and so in some cases the powder is given a reduction treatment.

Oil atomization

To overcome the surface oxidation with alloy steels a process has recently been developed in Japan in which paraffin (kerosene) is substituted for water, and such powders are now on the market.

Gas atomization

Although water atomization accounts for the bulk of atomized powder produced for pressing, there are several uses where irregular powder is either unnecessary or actually undesirable. Some of these will be discussed later. Then gas atomization may be used. The principles are similar to those for water atomization but jets of gas are used. It hardly needs to be said that the cooling rates obtaining are much lower and the droplets have time to assume a more or less spherical shape. The longer cooling period necessitates the use of larger atomizing chambers in order to ensure the extended free flight path of the droplets. The process is usually carried out vertically downwards, but it can be done horizontally or even vertically upwards which latter method is normal for aluminium. The slower cooling results in much more regular shaped particles, but the shape depends also on the surface tension of the metal. For example, gas-atomized brass powder is nowhere near spherical.

In theory any gas may be used; air is used for aluminium and nitrogen is used where oxidation is a problem. If freedom from nitride is an essential requirement, as in the case of the special nickel- and cobalt-based superalloys for aerospace applications, argon is used. Even helium, expensive though it is, has been used because of the more rapid cooling that it provides.

Centrifugal atomization

There are several circumstances in which even inert gas atomization does not provide an ideal solution. For example, molten metal melted in a crucible and passed through the nozzle of a tundish is liable, nay almost sure, to pick up refractory particles to some extent, and these unavoidably appear as non-metallic inclusions in the final powder, adversely affecting the properties of the eventual consolidated metal, especially the fatigue properties. Methods are available for 'cleaning' such powders, i.e. removing refractory particles, but the separation is not complete; metal particles with adhering or included refractory obviously present a problem. Another factor is that with argon-atomized powders there are always some hollow particles containing gas. As the argon does not react with the metal, nor can it escape by diffusion, there will be discontinuities in the finally densified material with consequent lowering of properties. The amount of included argon is much smaller in fine particles and manufacturers claim that by removing the coarser fraction the problem is reduced to negligible proportions.

However, both the above problems are largely eliminated by centrifugal atomization. As the name implies, droplets of molten metal are flung by centrifugal force from a reservoir of molten metal which can be produced out of contact with refractory of any kind, for example by arc melting in a horizontally rotating water-cooled metal hearth. Many different systems have been developed, including one in which the metal is melted by an electron beam or plasma arc and thus can operate in a near vacuum. In one such process a horizontally mounted rotating billet is used to supply the metal, a plasma arc melting the free end of the billet. These relatively expensive processes are viable only for powders for the most exacting applications or for powders of highly reactive metals; and complete 'cleanliness' of the powder is not guaranteed, it depends on the extent to which the starting material is free from non-metallic inclusions.

The whole subject of powder production has developed markedly in the last decade and there is good reason to believe that further significant changes are on the way.

3 Powder Properties, Characterization, and Mixing

In the production of PM components, the properties achieved are to a considerable extent a function of the properties of the starting materials, i.e. the individual powders and the powder mix. The relevant features can best be understood by consideration of the tests that are routinely specified.

Chemical

The basic chemical composition of the powder-based component has, of course, the same role in determining the mechanical properties, corrosion resistance, electrical and magnetic properties, etc as it does in parts made by other shaping processes, and no more need be said on that score. However, an additional chemical factor is the nature of the surfaces of the particles, especially the extent to which they are oxidized. This latter can for some metals be expressed as a *hydrogen loss value*, i.e. the loss in weight that occurs when a weighed sample is heated to a temperature below the melting point under standard conditions in a stream of hydrogen. This value is normally expressed as a percentage. The test does not reduce stable oxides such as those of aluminium, chromium, and the like, nor does it remove the oxygen present as oxide inclusions within the particles. The hydrogen loss test is, therefore, inapplicable to the powder of any metal the oxide of which is not reduced in nominally pure hydrogen, and it is relevant mainly to iron and steel and to copper and its alloy bronze. It cannot be used for brass because of the complication caused by volatilization of zinc, and the results are falsified if the original powder is contaminated by water or a volatile organic substance. More elaborate chemical tests which include melting of the powder are available for determining the *total oxygen* in a powder, but the hydrogen loss test is, for many purposes, a more useful factor, and is, moreover, easy and quick to perform.

With ferrous powders—iron and steel—another chemical test is sometimes specified, that is the determination of the percentage of *acid-insolubles*. A weighed quantity of powder is dissolved in a mineral acid and the residue is weighed. This consists mainly of refractory particles, silica for example, and they are relevant to the wear on the compacting tools and to the inclusions that will be present in the finished product. The latter is of importance only when we are concerned with fully dense products; with run-of-the-mill sintered parts, the pores present are of far greater significance in their effect on impact and fatigue properties, which are the properties mainly affected by non-metallic inclusions.

Particle Shape and Powder Density

Although shape is of considerable importance, it is not normally specified or determined directly. It is not an easy factor to measure and almost impossible to quantify. ISO standard 3252 illustrates a number of commonly occurring shapes under the names acicular, granular, irregular, flakey, spheroidal, and others. However, particle shape has a major influence on the packing density of powders, and this is a factor that can be measured. Spherical powders pack most densely, and the density decreases as the irregularity of the particle shape increases. There are two ways of expressing this bulk density that are in general use in PM. The most widely used is *apparent density* (AD) which is the density of a powder mass expressed as grammes per cubic centimetre, of a standard volume of powder gently loaded under specified conditions into a cylindrical container of specified volume. In the other test a weighed sample is placed in a graduated cylinder which is tapped or otherwise vibrated in a standard way until the volume is constant. This gives the *tap density* (or *tapped density*) which is, of course, higher than the apparent density. Density, particularly constancy of density, is relevant to die filling, it being fundamental that each compact has substantially the same weight of powder. Another important factor in connection with die fill is the ease with which the powder flows. Conveniently, the same equipment is used for flow and AD testing.

The *flow factor* or *flow time* is measured by timing the flow of a specified weight of powder from a standard funnel having a specified orifice. The term flow rate, which is sometimes used, is to be discouraged since rate would be expected to be expressed by a term such as grammes per minute which is not the case. Flow is always expressed as a time, which is, of course, the reciprocal of a rate. The most widely used equipment is the *Hall Flowmeter* illustrated in figures 7 and 8. The weight of powder used is 50 g and it overflows the cylindrical cup. The surplus standing above the rim of the cup is carefully removed by drawing a straight edge, say the edge of a steel ruler, across it, avoiding any vibration. The sample, the volume of which is 25 ml, is then tipped out and weighed.

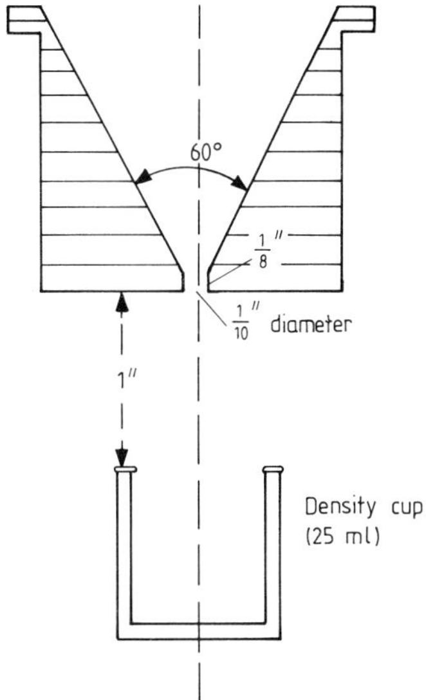

Figure 7 Schematic representation of the Hall Flowmeter.

It may be deduced that, all else being equal, flow time will be inversely related to AD, the volume of the standard weight of

Figure 8 Hall Flowmeter in use. *Norddeutsche Affinerie*.

powder used in the flow test being greater and, therefore, taking longer to flow through the orifice the lower the AD. However, all else is not always equal; an important factor affecting the flow and to some extent the AD is the size and size distribution of the powder particles. It is well known that very fine powders of any sort flow with great difficulty or not at all. Particle size and size distribution affect not only flow but very importantly the behaviour during sintering; fine powders sinter much more readily than coarse powders by reason of their greater potential energy associated with the larger surface area.

Particle Size and Size Distribution

Except in the case of spherical powders, the expression of size as a diameter, for example, is of doubtful validity, the more so the greater the degree of asymmetry of the particle. Are we to use a figure representing the maximum linear dimension or should we take an average? Is an acicular particle 1 mm long by 0.05 mm diameter greater or less than an equiaxed particle 0.5 mm across? Many different systems and appropriate test methods have been proposed and some are in use for special purposes, but for the bulk powders with which we are mainly concerned in PM we leave aside the finer points and rely on sieving. What exactly is determined by this process need not concern us greatly so long as we can ensure that the results we get are reproducible and relevant. Basically, sieve size refers to the largest sphere that will pass through its orifices, which are in fact approximately square in the case of the widely used woven wire screens. When, therefore, we give a figure for particle size it is this diameter to which we are referring. Size is usually expressed in microns (μm). The standard test is to put a weighed sample of powder on the topmost of a stack of circular wire mesh screens of decreasing aperture from top to bottom, shake and tap the assembly on a standard machine for a given time designed closely to approach the end point, and then to weigh the separate fractions or cuts as they are often called. The results are expressed numerically as the percentage of the total retained on each screen, or in the form of a histogram or sometimes as a smooth curve. A typical size distribution for a pressing grade of iron powder is given in table 1.

Table 1 Typical particle size distribution for pressing grade powders.

Fraction (μm)		% by weight
	>180	trace
<180	>150	0.2
<150	>105	21.6
<105	>75	25.9
<75	>63	11.8
<63	>45	14.3
<45		26.2

In the past, wire mesh screens were classified according to the number of strands of wire per linear inch, and so powders were referred to as, e.g. minus 100 mesh, meaning that substantially all the particles would pass through a 100 mesh screen. A moment's reflection, however, will show that this method is unsatisfactory because the actual orifices will vary according to the diameter of the wire used. Nevertheless, reference to mesh numbers will still be encountered in relation to powder size.

A word of caution is in order in connection with sieving. No two screens are absolutely identical, they become worn and they stretch in use, the orifices thus increasing in size. So, on an identical sample of powder two equally competent operators using nominally identical equipment can report marginally different results. The determination of particle size distribution on mixes is of limited value since the particle sizes of the several ingredients may be very different. For example copper powder is normally minus 150 μm while the tin powder with which it is mixed to make bronze is commonly all minus about 20 μm. Comparable differences exist with mixtures of iron and nickel.

Sub-sieve sizing

With very fine powders classification by sieving is impracticable or even impossible, and methods using quite different principles have to be employed. There are several processes available, one in fairly common use being to determine the surface area of a weighed quantity of powder. If two powder samples have similar morphologies, i.e. particle shape, the finer powder will have the larger surface area. It will be clear, however, that particle size as defined earlier is not the only factor; a highly irregular particle will have a much greater surface area than a spherical particle of the same included volume. For this reason, surface area measurements are of real significance only when comparing samples of similar powders made by the same process. One well known apparatus working on this principle is the Fisher sub-sieve sizer.

Compressibility

This term and the synonymous *compactability* refer, as will be self-evident, to the ease with which a powder mix can be compacted under pressure, but in powder metallurgy the meaning is more specific and refers to the density of the compact that can be achieved with a given compacting pressure. Compressibility is normally quantified as the density achieved by the application of a specified pressure, e.g. 100 MPa, but it can also be reported in the form of a compressibility curve showing density as a function of the applied pressure. It might be supposed that a softer powder will have a higher compressibility and up to a point this is so; but since in the standard test the starting point is a given weight of powder rather than a given volume, the apparent density of the powder also comes into the equation. The so-called high compressibility iron powders owe their better compressibility to the fact that they have a higher density to start with. Another factor related to compaction is the *compression ratio*, that is the ratio of the volume of the powder to that of the compact made from it. This factor is of importance in relation to green strength.

Green Strength

A PM part in the unsintered condition, i.e. as-compacted, is referred to as 'green'. The term green compact which is often encountered is tautologous since the term 'compact' is defined as a part in the as-compacted condition. The term '*green density*' is thus self-explanatory as is the term *green strength*. The latter is of importance only in connection with the safe handling of the compacts between the press and the sintering furnace, but that is, of course, of considerable importance. Strength may be measured in a number of ways. A widely used method for porous bearings is to fracture a hollow cylinder—a ring—in radial compression. This gives the *radial crushing strength* expressed as a K-factor, given by the equation

$$K = \frac{P(D - T)}{LT^2}$$

where P is the load, D the outside diameter of the specimen, T the wall thickness, and L the length. For ferrous parts a frequently used alternative is a bend test for which a strip of rectangular cross section is used. This gives a figure referred to as the *transverse rupture strength*, which is given by the formula

$$\frac{3PL}{2ba^2}$$

where P is the breaking load in newtons, L is the distance between the supports in millimetres, b is the width of the test piece in millimetres, and a the thickness in millimetres. The units are, therefore, N mm^{-2}. Since, in some cases, crumbling of the edges and corners of the compacts is at least as likely as actual breakage if not more so, some manufacturers specify a kind of tumbling test as a measure of resistance to that kind of damage. One such is the Rattler Test.

The origin of green strength has been much debated. It has been suggested that, on compaction, the surface irregularities of the particles get hooked together—the so-called jigsaw effect—but it is now largely accepted that the principal, if not the only, source of green strength is the formation of true metallurgical bonds, i.e. cold welds between the particles at the areas of contact. At these points distortion occurs and metallurgically clean surfaces are exposed which literally weld together. This occurs even with metals such as aluminium that have a surface layer of oxide which prevents welding, but this oxide skin is not ductile and, therefore, is ruptured by the distortion. It follows, from the above, that the greater the compression ratio for a given powder, the higher will be the green strength, which explains why powders that owe their high compressibility to higher initial AD give lower green strength at any given green density. It will be obvious also that the green strength will be influenced by the state of the surfaces of the powder particles. Surface films, be they of oxide, oil, or admixed lubricant, will lower the green strength, often very significantly.

It has been mentioned that green strength is desirable principally to ensure the safe handling of the compacts, and so the minimum green strength necessary depends on how carefully the parts manufacturer is prepared to handle them. Specifications vary between about 250 N mm^{-2} and 1000 N mm^{-2} or more.

Blending and Mixing

These two terms tend to be used somewhat indiscriminately but they have different meanings. In the ISO standard, blending is defined as the thorough intermingling of powders of the same nominal composition. This is done in order to produce a batch of powder that is uniform throughout as regards composition and particle size distribution. Mixing is defined as the thorough intermingling of powders of two or more different materials. Thus, we make a blend of several separate batches of, say, atomized copper powder, but we make a mix of copper powder with tin powder to make bronze, and of iron with copper, nickel, and/or molybdenum. In order to test for compressibility and green strength, even of individual powders, it is necessary to mix the powder with lubricant—see chapter 4 on compaction. Because, as a rule, the same equipment is used for blending and for mixing, it is called indiscriminately a blender or a mixer.

Mixers

Mixing is, of course, required in a wide range of industrial processes and many widely different designs of mixer are on the market differing not only in detail but also in the principle used. For metal powders the most widely used types are tumble mixers typified by the double-cone variety illustrated in figure 9. An essential requirement is that the powder rolls and folds over itself. It should not at any stage fall freely through the air as this, instead of causing mixing, can actually lead to segregation, especially if the mix contains ingredients of very different specific gravities, e.g. graphite and iron. A second type in common use in the ribbon blade mixer, more commonly called a ribbon blade blender, perhaps because of the alliteration. This type is generally a trough-shaped vessel, the bottom of which is of circular cross section, and in it, with only a small clearance, rotate helically disposed blades mounted on a horizontal shaft.

Time of mixing is of great importance. Obviously, a certain time is required to produce a homogeneous mix—it may be as low as 2 minutes—but to continue tumbling after that point is not only a waste of time, it can be positively damaging. The tumbling action inevitably involves mechanical work being applied to the particles and this, by flattening the asperities, increases the AD of the mix which has *per se* a detrimental effect on green strength. Furthermore, the more complete coating of the particle surfaces with

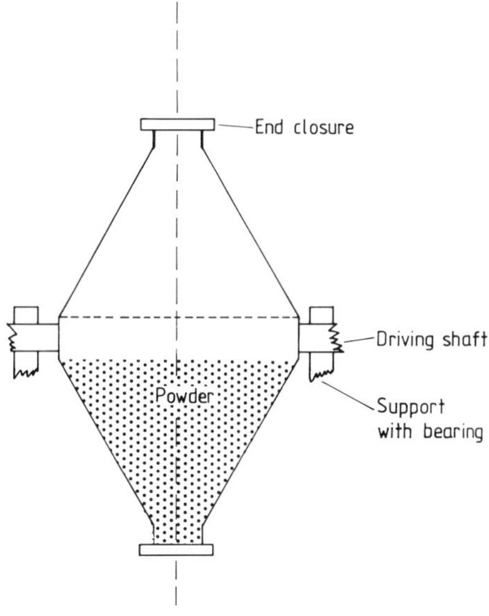

Figure 9 Section of double-cone blender/mixer. The vessel is caused to rotate end over end at a speed usually between 15 and 30 rpm so that the powder rolls over and over. Such blenders are made in capacities from 5 kg for laboratory use up to as high as 50 tonnes for blending ferrous powders.

lubricant damages the flow properties of the mix and also contributes to the further reduction of green strength.

Handling and storing

Having achieved a homogeneous mix *in the mixer*, precautions are needed to ensure that segregation (or de-mixing) does not occur between the mixer and the die cavity in which the mix will be compacted. Segregation may occur as the powder is run out of the mixer, during transport of the containers, and, very importantly, as the powder is led from the press hopper to the die. De-mixing is more likely when the mix contains powders of widely different particle sizes or of widely differing specific gravity. Mention was made earlier of the case of iron/graphite mixes, but the danger exists also with mixes of iron with very fine nickel or molybdenum powder, and with mixes of relatively coarse copper and fine tin. The added alloying elements in iron are required to be fine in order to facilitate rapid alloying during sintering.

Preventing segregation

A big step forward has been the introduction of what the manufacturers call partially pre-alloyed powders, pioneered by Höganäs in Sweden. The powder mix is spread on a steel band and passed through a furnace at a temperature sufficient to cause some interdiffusion of the ingredients but insufficient to sinter the powder. The fragile cake is then crushed and re-sieved. A still more recent development, designed to minimize the segregation of graphite, is the inclusion in the final mix of an organic ingredient that basically acts as a glue, causing the graphite particles to adhere to the iron.

4 Compaction

Compaction of the powder is, arguably, the most important single step in the whole PM process. It has been mentioned already that the economic viability of the process depends critically on its ability to produce components to the required shape and with dimensions after sintering so close to those required that little or no machining is needed. The achievement of this essential feature is largely determined at the compacting stage. The basic dimensions of the part are, of course, determined by the geometry of the tools and the accuracy of their setting on the press, but the dimensions after sintering depend also on the changes that take place in the sintering furnace, and these are influenced by the density and the uniformity of the density of the compact. This uniformity is a function of tool design and of the motions of the press, and also of the constancy of the manner in which the powder is fed into the die cavity. Rigid dies are used for the vast majority of sintered parts, and it is to this process that the following discussion applies.

At its simplest the tools required are a block of metal—the die—with a vertical sided hole in it—the die cavity—and a rod of the same cross section as the cavity—the punch. Normally two punches are used, an upper punch which descends into the die cavity and a lower punch which closes the bottom of the cavity and serves to push the compact from the die (see figure 10).

Leaving aside purposely porous parts which are the subject of a separate chapter, the final density of structural sintered parts is always and inevitably less than the theoretical density of the metal or alloy concerned, which is another way of saying that PM parts are porous. From this arises the concept of *density ratio*, the ratio of the actual to the theoretical. It is normally expressed as a percentage and values between about 70 and 90% are common.

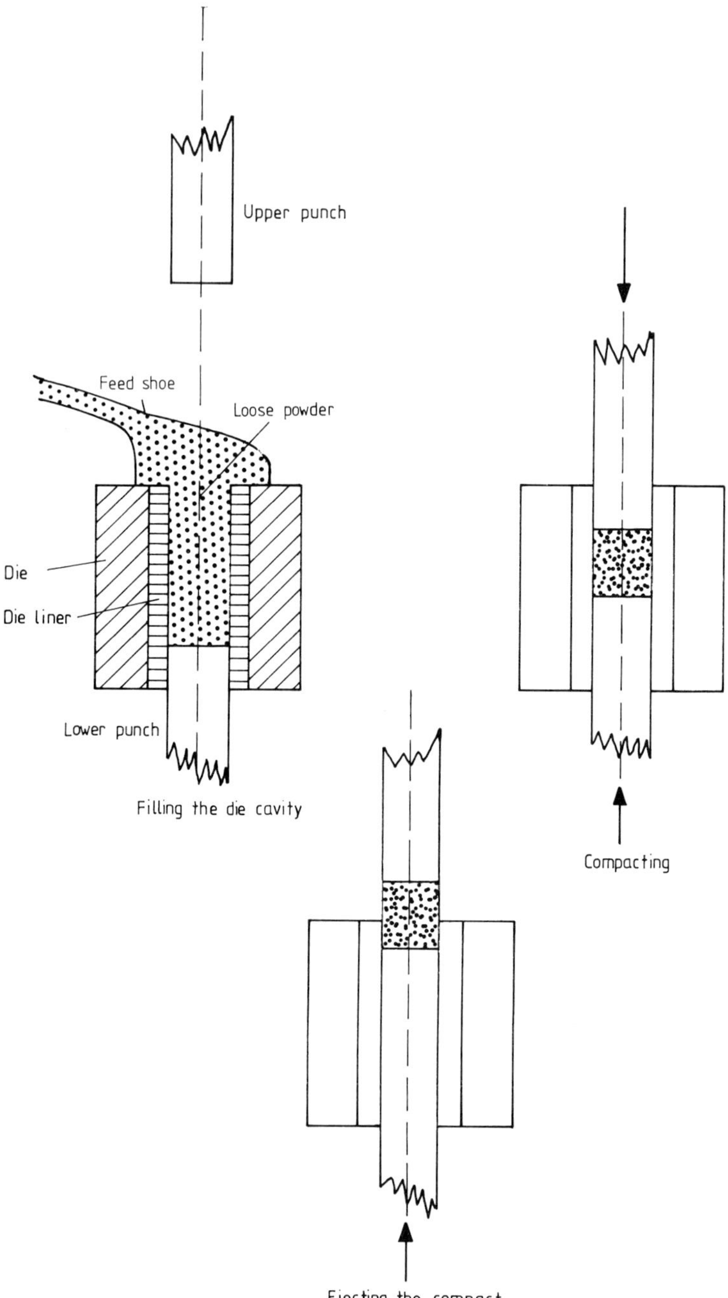

Figure 10 Production of a compact.

It need hardly be said that the higher the density ratio the stronger the part will be but it will also be more expensive to produce, and for many applications a lower strength is not a disqualification. Nevertheless, increasing emphasis is being put on strength as powder metallurgy is continually challenging other metal shaping processes for critical load bearing applications. Increased strength can be obtained by changes in composition, e.g. by using alloy steels, and by appropriate heat treatment, but even then higher density will still further improve the strength, as well, incidentally, as other mechanical properties, especially fatigue and impact strength. We shall see later that compact density is important also to the sintering process itself, and in order to end up with accurate dimensions after sintering uniform compact density is desirable. This is not easy to achieve. If we consider a powder mass in a die, and assume the loose density to be uniform, we shall need to apply uniform pressure *throughout* the powder mass in order to get uniform density throughout the compact. But we are applying the pressure in one direction only, and since powders do not behave like liquids which transmit an applied pressure uniformly, the pressure on the powder mass will decrease as the distance from the punch through which it is applied increases. Assuming that we are pressing from one end only, usually the top, the pressure and, therefore, the compact density will decrease from top to bottom of the compact. Matters can be improved by pressing from both ends. A comparable result is obtained by allowing the die to move downwards relative to a stationary lower punch under the frictional force on the die wall, this being called the *floating die system*. Even then, the density in the middle of the compact will be lower than that at the ends, and one consequence of this will be variation in the dimensional change on sintering—the less dense regions shrinking relative to regions of higher green density (see figure 11). In practice this feature limits the length of compact that can usefully be made; a length-to-diameter ratio of 3 to 1 is often quoted as the maximum.

When we come to more complicated shapes variation in density becomes even more significant and of such importance that one of the PM companies has developed an instrument capable of measuring density locally. This instrument uses the principle of the absorption of gamma radiation and is called the Gamma-Densomat. The reduction in density as the distance from the punch(es)

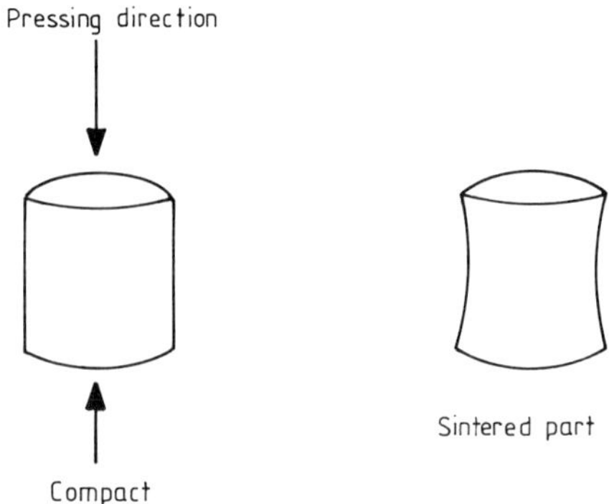

Figure 11 Non-uniform shrinkage caused by density variation (greatly exaggerated).

increases is largely accounted for by *die wall friction*, i.e. friction between the powder mass and the vertical faces of the tools—die walls and core rods—hence the importance of lubrication. It is almost universal to provide lubrication by mixing an organic lubricant with the powder before delivering it to the press, the most common being metal stearates, usually zinc stearate, or a wholly organic wax-like substance, and sometimes a mixture of the two. The lubricant, normally between 0.75 and 1% by weight, occupies a significant volume and, therefore, places an absolute limit on the density that can be achieved. The usable density is, however, limited by the necessity to allow the lubricant to escape during the heating-up process; and *de-waxing* is a major preoccupation in the design of sintering furnaces and of the sintering regime generally. A further feature of lubricants is that they reduce the flowability of powder mixes and tend to lower the green strength of the compacts. For these reasons, a lot of effort has been put into methods of applying a lubricant only to the tool surfaces, and successful *die-wall lubrication* processes have been reported. The problems, however, are formidable, nevertheless commercial use is being made of the process. It should be mentioned that some workers argue that admixed lubricant has a further function, that is to facilitate the sliding of the metal particles relative to each other, thus assisting the attainment of uniform density in the compact, but convincing evidence of this is lacking. It is, moreover, arguable that sliding of the particles over each other is not desirable since it would inhibit the metal-to-metal contacts on which the development of green strength depends. From the above it follows that increasing the compacting pressure will give increased green strength because there will be greater plastic deformation of the particles at the points of contact and thus a greater area of metallurgical bonding. The relation between green strength and density is shown in figure 12.

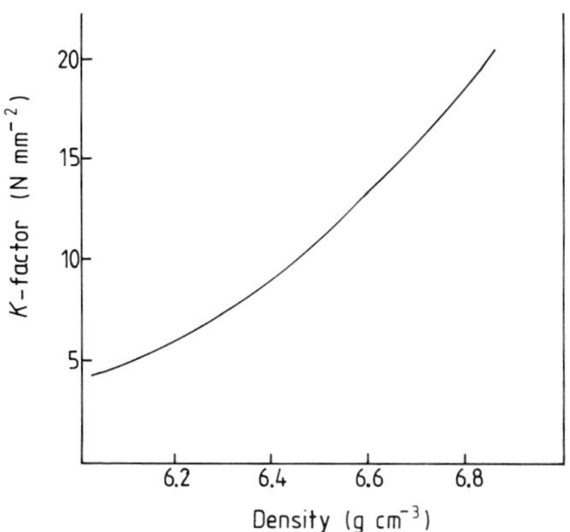

Figure 12 The relationship between green strength and density (Radial Crushing Test).

Compact Size

At first sight it might be supposed that the compact will have the same cross section as the die cavity but this is not so. We have referred to rigid dies, but in practice any die of finite dimensions will undergo some lateral expansion during the compaction as a result of the internal pressure and, of course, this expansion increases as the applied pressure increases. This means that the compact when in the die is larger than the nominal die size and, additionally, it is under compressive stress; so, when the compact is ejected from the die and this compressive stress is released, the compact expands slightly. This expansion, referred to as *spring back*, results in the compact having a greater cross section than the die cavity. Because the extent of this difference varies, it is normal when discussing dimensional aspects, e.g. changes of dimensions on sintering, to refer to die size rather than to the actual size of the compact since, for a given part, die size is the one fixed point. The above relates to the dimensions at right angles to the pressing direction. When we come to consider dimensions parallel to the pressing direction we need to know what type of press is being used. There are fundamentally two different types: mechanical and hydraulic. The former, operated by cam or toggle action, gives a constant stroke to the punches, thus producing compacts that are to a first approximation always the same height, and if the amount of powder in the die, that is the fill, varies, the compact density will vary. Hydraulic presses, however, are generally programmed to apply a fixed pressure, so that for a given powder mix, compact density will be constant but the height will vary if the fill, i.e. the weight of powder in the die, varies. In some cases, however, mechanical stops are incorporated to limit the movement of the punches and thus give constant height to the compacts. The importance of variation in height depends on the part being made. In many cases a slight variation in height may be of little consequence, but in some circumstances, for example where the part is to be subsequently forged to high density in a closed die, a significant variation in height could be a complete disqualification. The achievement of constant die fill is, therefore, a desirable objective, and to that end the apparent density of the powder used must be consistent. With good mixing procedure AD will be uniform throughout a single batch, but it is not possible in industrial practice to guarantee that each batch will have exactly the same AD. This means that adjustment to the depth of the die cavity may be required when a new mix of powder is to be used.

The above considerations have been illustrated by reference to the pressing of simple shapes, but a glance at the illustrations will show that in practice most sintered parts are quite complicated, having several levels, irregular cross sections, and through-holes. Such parts require complicated tool assemblies with multiple punches and core rods (see figure 13). As can be imagined, the design of the tools and of the press movements, to produce such compacts and to give them uniform density, is a very skilled job.

The tool setter also is a key operator in the PM plant. We shall see in the next chapter how variations in density throughout the compact can lead to distortion on sintering.

Figure 13 Tool set for compaction of the valve rod guide for a Ford engine. The die cavity shows the general shape of the cross section which measures 8 cm across the longest diagonal. Note the three core rods for the through-holes which can be seen on the two main punches. The two subsidiary punches take care of the leg sections, one of which is 6 mm thick and the other 5.5 mm. The thickness of the rest of the part is about 3 mm. *Pankhurst Tool and Die Co Ltd.*

Tool Materials

The choice of materials for dies, punches, and core rods depends very significantly on the number of identical parts that are to be made. At the lower end alloy steels may be used, but more wear-resistant materials such as high chromium steels, high speed steels and, ultimately, hardmetal (cemented carbide) are preferred for long run production. Die life varies considerably: it depends not only on the die material but also on the surface finish and, importantly, on the nature of the powder being pressed, the compact density and the lubrication provided. In this connection the acid-insoluble constituents in iron powder already referred to are relevant. With steel dies, up to about 200 000 compacts may be produced, and with carbide dies, 1 000 000 parts or more are possible. The longer life and, of equal importance, the saving of down-time on the press while the setter fits new tools fully justify the use of the progressively more expensive die materials. Moving parts such as core rods and punches are easier to replace and it is not uncommon to use carbide dies and alloy steel for punches and core rods. Apart from the initial cost aspect, steel has better ductility and impact strength than hardmetal and is, therefore, less liable to fracture.

Design of Sintered Parts

The general shape of any engineering component is, of course, dictated by the function that it is required to perform, but considerable variation in detail is possible without in any way prejudicing its performance in service. No matter what shaping process is used, the details of the design have to be tailored to the limitations of the process. This is not the place for a complete PM design manual, but three important considerations may usefully be mentioned.

(1) We have already seen that a fundamental requirement is that the compact must be capable of being ejected safely from the die. It is not possible, therefore, to mould into the compact grooves, projections, or holes except they be parallel to the pressing direction. If such features are required, they have to be produced by machining after sintering and sizing.

(2) A second consideration is the necessity to avoid excessively fragile tool elements. Very thin sections are obviously a potential source of failure, as are bevels and the like which would require

feather edged punches. This problem can usually be avoided by having a small flat at the edge. Very thin compact sections are undesirable for another reason; they introduce problems of poor powder flow into the die and non-uniform die filling.

(3) There are constraints also relating to the necessity to avoid cracks occurring when the compact is ejected. The ejection process itself and the associated spring-back to which reference has already been made, introduce stresses which can cause cracks if the designer has not taken them into consideration. Abrupt changes of section and sharp re-entrant corners are dangerous features than can usually be avoided, e.g. corners can be radiused—this is good engineering practice in any case.

Production Rates

The number of parts produced per unit time varies considerably, notably with the size and complexity of the part. With simple parts rates as high as one part per second can be reached using mechanical presses. Hydraulic presses which are more usual for large parts are generally slower, and 10 parts per minute is a good rate of production for large parts, even of relatively simple shape.

Overcoming Design Constraints

The limitations imposed by the necessity to eject the compact from the die can be mitigated in a number of ways.

(1) A *split-die* process developed by Olivetti and, therefore, often referred to as the Olivetti process, uses in effect two dies one above the other, powder being compacted from both ends. After compaction the two dies are separated and the compact extracted from the gap between them. The principle can be understood by reference to the simple shape shown in figure 14. Figure 15 shows a collection of parts made by this process.

(2) A similar end result can be achieved by making the part as two separate compacts and sintering them together one on top of the other, in which case some feature must be incorporated

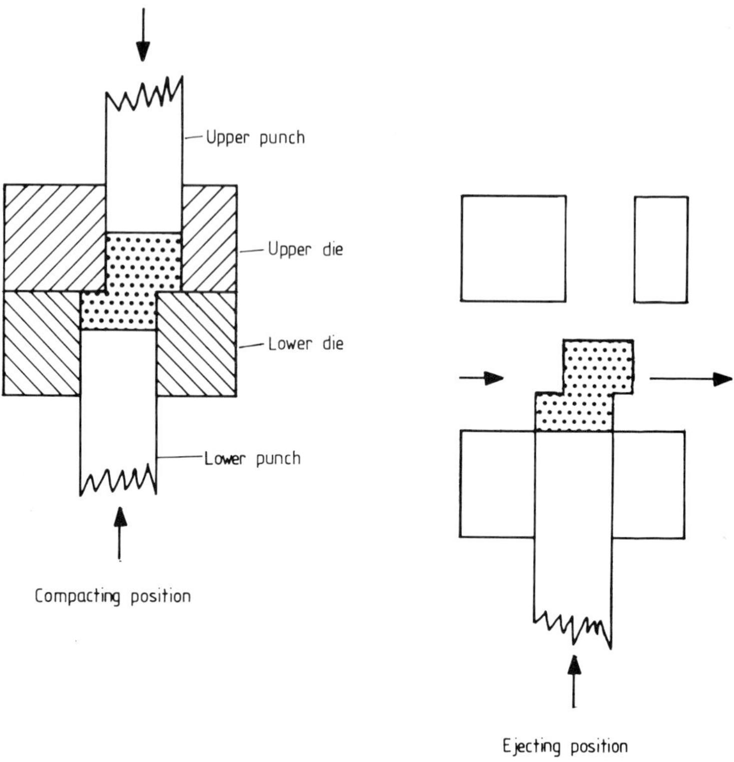

Figure 14 Schematic arrangement for split-die compaction.

in the design to ensure accurate positioning. Alternatively the two parts may be joined after sintering by projection welding or by brazing—see also the later section on infiltration in chapter 6.

Hot Pressing

Although the bulk of sintered parts is made by pressing the powder mix at ambient temperature followed by sintering, hot pressing is used in certain cases. At elevated temperatures metals are softer and, therefore, it is usually possible to press to much higher density

Figure 15 A collection of parts made by the split-die process. *Tecsinter*.

without increasing the pressure required. It may be possible to dispense with a separate sintering operation, but this is not general because hot pressing is only justified by the significantly better properties that are obtainable. A subsequent sintering step almost invariably improves the mechanical properties. The use of hot pressing is limited by the much greater cost; special heat-resistant dies are required, a controlled protective atmosphere may be necessary and production rates are relatively very low. The process is used in the production of hardmetal shapes and bonded diamond cutting tools, both of which are in any case expensive materials.

Other Powder Shaping Processes

Injection moulding

This relatively new process does not strictly fall into the category of compaction since little or no external pressure is involved. However, it fits in here in so far as shaping in a rigid mould takes place. Metal powder is mixed with a proprietary organic binder to form a plastic mass which is warmed and injected under pressure into a suitable mould that can be readily opened to remove the shaped part. A machine similar to those in common use for the injection moulding of plastics is used. The compact, if such it can be called, is then carefully heated to remove the binder and eventually sintered. The shrinkage that takes place is very large—some 20–25% linear—and this might be thought to militate against the achievement of good dimensional accuracy. However, in fact, the process can be so controlled that the shrinkage is consistent and predictable. Final densities of 95% or better are obtainable. A major attraction of the process is that many of the restrictions as to part shape that apply to the traditional compacting process are avoided. On the debit side is the fact that it is necessary to use very fine powders—below about 20 μm—which with a few exceptions are not those normally available on a commercial scale. Nickel was the first metal to be used for injection moulding, perhaps because fine nickel powder was already in regular production by the carbonyl process, and nickel and its alloys are still the most important compositions being made by injection moulding. Powders for cemented carbide manufacture are also in the right size range, and hardmetal parts also are being produced on a commercial scale. The success of the process is, moreover, acting as a spur to the development of processes to produce other metals such as pre-alloyed powders economically in the required fine state. Currently, the parts being made are small and are used in high technology applications, mainly aerospace applications, and, as might be deduced, they are expensive.

Cold iso-static pressing (CIP)

In this process the powder is fed into a mould made of an elastic material—polyurethane is highly favoured—and the whole is then immersed in a liquid, usually water, which is then pressurized. As the name implies, pressure is applied equally in all directions, and as there is no question of die-wall friction, no lubricant is required so that near-theoretical density can be obtained. Because of the high compression ratio between the powder and the compact the latter is very much smaller than the original mould which, after release of the pressure, returns to its original size so that the compact can be removed without difficulty. The compact may then be sintered in the normal way except that, since no lubricant is present, de-waxing problems do not intrude. As in the injection moulding process, many of the shape restrictions that apply to rigid die compaction do not apply and threaded objects and complicated parts in high speed steel are among those being made on an industrial scale. Unlike the injection moulding process CIP does not normally give parts with very high dimensional accuracy.

Hot iso-static pressing (HIP)

This is dealt with in chapter 7.

5 Sintering

Correct sintering is of paramount importance to the PM process, to ensure not only the development of the strength needed for the part to fulfil its intended role as an engineering component, but also that the dimensions of the part are correct. Without exception sintering requires heat, and the ISO definition reads: 'The thermal treatment of a powder or compact at a temperature below the melting point of the main constituent for the purpose of increasing its strength by bonding together of the particles.' Note that it says 'powder or compact', implying that uncompacted powder may be sintered and, as will be seen later, *loose powder sintering* is done purposely in certain special cases. We have seen already that in the manufacture of sponge iron powder the loose powder is partially sintered during its passage through the reduction furnace. However, our concern here is with the sintering of compacts. The definition above refers also to 'bonding together of the particles', and this implies the formation of true metallurgical bonds by crystal growth across the inter-particle boundaries. It has been said earlier that the formation of such bonds in the areas where neighbouring particles are deformed at their points of contact by the compacting pressure is responsible for the generation of green strength. During sintering these areas of metallurgical contact grow and the strength progressively increases.

Theory of Sintering

Theories about exactly what happens during sintering have provided the subject matter for innumerable conferences and learned scientific papers. There is in Belgrade an International Institute for the Science of Sintering, but we do not have the space nor, fortunately, is it necessary for our present purpose to go into the theories in any depth. Suffice to say that the enlargement of the areas of metallurgical contact takes place by atomic diffusion, whether it be bulk diffusion, surface diffusion, or both is debated. Research workers have started with the simplest case, that of spherical particles in contact, and it can be seen that, on heating, necks form between the particles at the point of contact and these progressively thicken. As a consequence, the originally sharp angles between particles are rounded and the spaces eventually become, in the ultimate, spherical pores. Thermodynamical considerations show that reduction of the area of free surface is the goal. The phenomenon of neck growth and pore rounding is illustrated in figure 16.

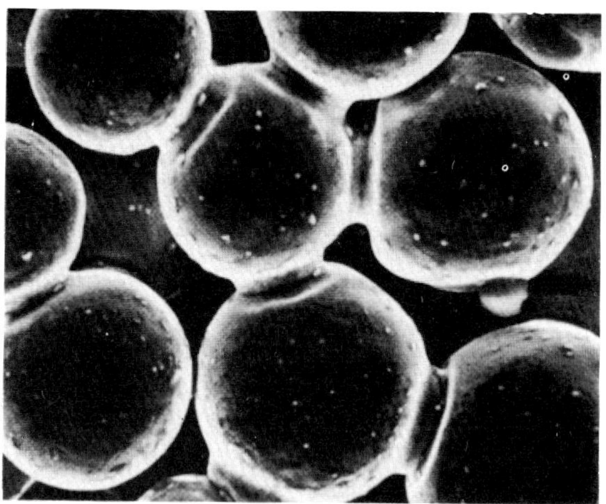

Figure 16 Neck formation between adjacent particles. These are of spherical bronze (magnification ×1600).

Eventually the outlines of the original particles disappear as recrystallization and grain growth occur across the boundaries, but these processes may be hindered if the particles are covered by a film of oxide or other, extraneous, matter. Impurities concentrated at *prior particle boundaries* is a well known defect.

Sintering Practice

For the most part, sintering is done in continuous furnaces having woven wire mesh belts on which the parts are conveyed. In many cases the parts are loaded directly on the belt, but in certain cases they are put into trays or on refractory ceramic tiles. During the initial heating-up phase, or pre-heating as it is sometimes called, the organic lubricant is expelled. There are two important requirements: firstly, the lubricant should be expelled as vapour without decomposition, as decomposition would leave a carbonaceous residue; secondly, the rate of heating should be such that the lubricant is driven out from the whole section. Obviously, the surface layers of the compact are heated first and the heat penetrates gradually into the centre. If the heating rate is too great it can happen that sintering of the surface layers with pore closure takes place before the lubricant in the centre has evaporated so that its escape route is blocked and blistering may follow. A too rapid heating rate can lead also to distortion caused by temperature differentials between thin and thick sections. It is standard practice, therefore, for sintering furnaces to have a separately controlled *de-waxing* zone at a lower temperature than that of the main sintering zone. Apropos of de-waxing, it is further required that a sufficient flow of gas be maintained to sweep away the vapour which might otherwise condense on and disfigure the surfaces of the incoming parts.

After passing through the de-waxing zone, the parts enter the main sintering section, which itself may consist of a number of separately controlled zones in order to ensure as constant a temperature as possible during the time the parts are in the sintering section. On leaving this section the parts enter the cooling zone, which is normally surrounded by a water jacket, and eventually emerge from the furnace at a temperature low enough at least to avoid oxidation on contact with the air. The cooling section is often equal in length to the whole of the rest of the furnace. A typical temperature profile is shown in figure 17.

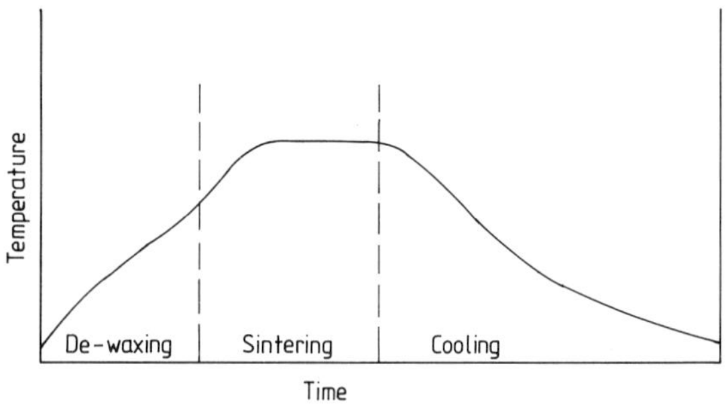

Figure 17 Temperature as a function of time during sintering.

Furnace atmospheres

For metals other than the so-called noble metals that either do not form stable oxides—gold, platinum and its relatives—or the oxide of which decomposes at a low temperature—silver—sintering has to be done in a protective atmosphere. This is almost invariably a reducing atmosphere because, apart from preventing oxidation of the metal being sintered, it may be required to reduce the unavoidable films of oxide on the surfaces of the particles. Atmospheres in common use include partially combusted hydrocarbons such as methane, butane, and propane, dissociated (cracked) ammonia, so-called synthetic atmospheres made by careful mixing of prescribed proportions of hydrogen and nitrogen, pure hydrogen and, if it can accurately be called an atmosphere, vacuum. Combusted hydrocarbons provide the cheapest form of protective atmosphere and for that reason alone are very widely used. By varying the air-to-gas ratio, a wide range of compositions is obtainable. Since the combusted gas contains water vapour it is desirable to dry it to a dew point below 0 °C for satisfactory operation with ferrous components. By using a very limited amount of air the atmosphere may contain up to 45% of hydrogen, some carbon monoxide and carbon dioxide, with nitrogen as the remainder. Because of the endo-thermic nature of this reaction, external heat has to be supplied, and for that reason the resulting atmosphere is called *endo-gas*. If the hydrocarbon is burnt with just insufficient air for complete combustion, an atmosphere which may contain as little as 5% of hydrogen and a very large percentage of nitrogen is produced, and as this reaction is exo-thermic, the atmosphere is called *exo-gas*. This is the cheapest sintering atmosphere available, but its reducing potential is low and thus the removal of oxides from the powder compacts is less efficient and lower strengths may result. For the sintering of steels, i.e. iron-based alloys containing carbon as an essential alloying element, the *carbon potential* of the atmosphere is important; it should be in equilibrium with the steel so that neither decarburization nor carburization takes place.

Synthetic- or, as they are also called, nitrogen-based atmospheres have become increasingly popular during the last decade because of their reliability, low dew point, and the ease with which the hydrogen content can be adjusted. Furthermore, with wider use in industry generally, the cost of the two gases has come down, as has that of the equipment necessary to handle and meter them. In fact, because of the enormous usage of oxygen in steel making, nitrogen can, in theory, be had almost for the taking, and the gas suppliers purport to have shown that these nitrogen-based atmospheres are not uncompetitive vis-à-vis endo-gas. For use with steels, the necessary carbon potential is provided by a controlled addition of hydrocarbon gas. In all cases the atmosphere is generated (or mixed) outside the furnace, and quite elaborate analytical facilities that send appropriate signals back to the generator are provided to ensure that the composition of the gas is kept constant and to specification.

Vacuum sintering

Vacuum can be regarded as a special case of a controlled atmosphere, and in many cases is probably the best from a scientific point of view. Gas within the compacts is readily extracted and even metals with stable oxides such as titanium, zirconium, and tantalum may be sintered. Vacuum is used also for hardmetals, especially those containing titanium or tantalum carbide, and for high speed steels. Vacuum sintering is, however, very expensive. Apart from equipment cost and relatively low production rates, it is a batch process difficult if not impossible to make continuous or to automate. For these reasons it is unlikely to find application in the production of standard, run-of-the-mill mechanical parts for which the atmospheres referred to earlier give entirely satisfactory results.

Furnace design

There is wide diversity in the details of the design of sintering furnaces and new and improved designs appear from the furnace manufacturers at quite short intervals. We need not be concerned with the details; it will be sufficient to present the general principles of the most widely used type—the continuous mesh belt furnace. Mention has been made already of the three sections: the pre-heating or de-waxing section, sometimes referred to as the burn-off section; the main sintering section; and the cooler.

These sections are joined in tandem, usually horizontally, but some users have favoured a hump-back construction in which the sintering section is horizontal, but the de-waxing and cooling sections descend from it at an angle such that the ends are below the level of the furnace hearth. The idea is that as the protective atmosphere is lighter than air, the risk of air diffusing into the furnace is minimized.

The de-waxing and sintering sections have steel cases lined with refractory and have heating elements, most commonly electrical resistance heaters, inside. They have heat resisting metal hearths. The cooling section is a rectangular sectioned steel tube surrounded by a water jacket. Some furnaces have a heat-resistant metal muffle inside the pre-heating and sintering zones to ensure gas tightness, and such muffles serve also towards ensuring temperature uniformity. Furnaces that do not have muffles must have a gas-tight outer casing. The conveyor belt rests on the bottom of the furnace chamber, which may be as much as 60 cm or more in width. The height of the furnace chamber is governed largely by the height of the parts for which it will be used. There must be sufficient clearance, or head-room as it is called, but beyond that, extra height is disadvantageous. Extra height increases the size of the furnace and introduces the danger of stratification of the atmosphere which could lead to inefficient contact between the reducing gas and the work load. At the entrance and exit to the furnace chamber, doors or curtains are fitted to act as rough seals, but ingress of air is largely prevented by a significant outflow of the protective gas atmosphere which is necessary anyway to sweep out the lubricant vapour. The gas is piped into the furnace usually at a number of points along and adjacent to the sintering section. Normally the outflowing gas is burnt as it emerges from the furnace, but in some modern designs it is burnt with a subsidiary air supply inside the de-waxing zone. The lubricant vapour also burns and provides much of the heat required for pre-heating.

The cost of the sintering operation is a very significant part of the total cost of producing sintered parts, and energy for heating, not only of the parts themselves but also of the belt and, very significantly, of the gas atmosphere, accounts for a large part of the cost. Many producers now make use of some of the waste heat by using the cooling water for space heating, but energy efficient designs are increasingly being installed, e.g. designs which allow the burning of the lubricant inside the pre-heating zone as mentioned above.

The mesh belts are normally of an 80/20 nickel/chromium alloy and they have a useful life when operated at temperatures up to 1150 °C. Whether it is coincidental that the normal sintering temperature for ferrous components is 1120/1130 °C, or whether that temperature was selected with an eye to making possible the use of such mesh belts is not clear. What is clear is that stronger components can be produced by using a higher temperature. For certain newer steel compositions higher temperature sintering is mandatory, and thus other types of furnace have to be used, together with different heating elements, because the usual element is, like the belt, made of nickel/chromium. Among the other types are pusher and walking beam furnaces.

A pusher furnace may be much the same as a mesh belt furnace in general design, but instead of conveying the parts on a moving belt they are loaded into trays which are pushed one after the other along the furnace hearth. Small-scale producers often start with this type of furnace which is, as could be expected, much cheaper to make. It requires an operator to push in a new tray at the prescribed interval of time and to remove and unload a tray at the exit end. Better end closures can be fitted than are feasible on mesh belt furnaces because it can be arranged that the doors, which can be fairly gas-tight, are opened only to admit or to remove the trays. Thus, the amount of protective gas may be lower. A variation of the pusher furnace is the roller hearth type in which the trays are propelled through the furnace on driven rolls.

In a walking beam furnace the parts are again conveyed in trays that are advanced at pre-determined intervals by a series of internal mechanisms that lifts them, moves them forward, and lowers them back onto the hearth, the lifting mechanism returning at a lower level to its initial position. The mechanism is made of refractory material and, therefore, such furnaces can operate at much higher temperatures than mesh belt furnaces—up to 1300 °C or more. Their use may well increase as the demand for higher sintering temperatures grows with the increasing use of special alloy steels.

Changes that Occur During Sintering

Theoretical considerations of what happens during sintering, although of great scientific interest, do not immediately concern us here, but we are vitally concerned with what effect these happenings have on the engineering properties of the parts. A gradual increase in strength has already been noted—this is the main purpose of the operation—but what happens to the shape of the part is only slightly less important. One of the laws of nature is that all systems tend towards the lowest energy state, and since free surfaces have energy associated with them, a reduction of surface area is the goal. This means that in porous materials the pores tend to become more nearly spherical and the object as a whole tends to shrink and so increase in density. This increase in density results in increased strength, but we can be less happy about the effect that this has on the dimensions. In practice, provided the powder mix and the die fill are constant, the change will always be the same and can be allowed for in the tool design. However, consistency is easier to talk about than to achieve, and some small variation must be accepted. Fortunately, small variations can be smoothed out by the sizing operation which is dealt with in chapter 6. As could be expected, dimensional consistency is easier to achieve if the dimensional change on sintering is small and, increasingly, parts producers are favouring mixes that give no dimensional change.

Control of shrinkage

Since shrinkage is, as we have seen, a natural consequence of sintering, the possibility of producing *no-growth mixes* requires explanation. The secret lies in the behaviour of metallic alloying elements added to the mix in elemental form and, in this context, the addition of copper to iron is of the greatest practical importance. About 2.5% of copper is sufficient to compensate for the shrinkage of iron compacts sintered in the normal way. Larger amounts cause actual growth. We say '*about* 2.5% of copper' because growth and shrinkage are affected also by other factors: the properties of the iron powder being used, its source, its exact composition and its particle size distribution; and also by the sintering regime—higher temperature and longer time tending towards shrinkage. The exact mechanism whereby copper causes growth is still the subject of debate. It was at first supposed that the expansion was caused by the diffusion of copper atoms into the iron lattice, but this view was no longer tenable when it was demonstrated that expansion takes place quite suddenly after the copper melts, while diffusion of one metal in another is quite slow. It has now been shown that when the copper melts it penetrates first along the boundaries between the iron particles and then along the actual grain boundaries of the iron, forcing them apart. The spaces originally occupied by the copper particles become pores. The whole process is, of course, dynamic, and if the compact be held at the sintering temperature for progressively longer times, shrinkage increasingly occurs.

The growth promoting effect of copper can be counteracted by the addition of nickel which, on its own, accelerates sintering, presumably by increasing diffusion rates. Thus we can produce no-growth mixes having different levels of strength. No mention has been made of the sintering atmosphere, but it cannot be ignored, and for reproducible results it too must be maintained constant.

A comparable growth controlling mechanism is available for bronze: if this alloy is made from mixes of elemental copper and tin powders, growth occurs on sintering. This important topic is dealt with in more detail in the section on porous bearings in chapter 8.

Other considerations in ferrous powder mixes

Copper steels have been mentioned at some length because of their interest from the dimensional point of view, but also because, being relatively easy to sinter, they are widely used. However, from the cost point of view copper and nickel are expensive alloying elements. More effective as a strengthening addition is carbon. Because of the importance of compressibility this element is almost always added in elemental form as graphite. Indeed there is advantage in using elemental powders in all cases, since pre-alloyed powders are inevitably harder and, therefore, require greater force for compaction, i.e. their compressibility is worse. In the case of carbon, the rate of diffusion in austenite, i.e. at temperatures above the α–γ transformation of the iron, is fast because it diffuses interstitially, and the material becomes homogeneous as far as carbon distribution is concerned. This is not the case with metallic alloying elements, as can be inferred from the above discussion on the effect of copper.

With carbon, however, there are some complications. At and even below the normal sintering temperature, carbon reacts with any oxide present on the surface of the iron particles, producing carbon monoxide and, of course, lowering the carbon content of the steel, as the material must now be called. Additionally, carbon, even when in solution in the iron, can react also with the sintering atmosphere. Among the chemical reactions that come into the picture are the following:

$$\begin{aligned} &C + FeO \rightleftharpoons Fe + CO \\ &(Fe,C) + 2H_2 \rightleftharpoons Fe + CH_4 \\ &(Fe,C) + CO_2 \rightleftharpoons Fe + 2CO \\ &(Fe,C) + H_2O \rightleftharpoons Fe + CO + H_2. \end{aligned}$$

All these reactions are what chemists call reversible reactions, i.e. they can go either way depending on which end you start from. Thus it can be seen that the eventual carbon content of the steel, at least in the surface layers, is dependent not only on the amount of graphite added, but also on the furnace atmosphere. Moisture, hydrogen, carbon monoxide, carbon dioxide and methane content are all significant, and control of the gas to ensure that it is neither carburizing nor de-carburizing is of the greatest importance. This leads us to the concept of *carbon potential* of the gas atmosphere, and the aim is to provide an atmosphere with a carbon potential that is in equilibrium with the intended composition of the steel. When we look very closely, we find that the question is even less straightforward than might be supposed; the equilibrium ratios of the various gases change with temperature. Fortunately, we do not need to be too exact, and with experience it is not too difficult so to control the atmosphere that satisfactory and reproducible results are obtained.

Liquid phase sintering

This refers to the process in which a liquid phase is present at some stage during the sintering cycle. The liquid phase is, of necessity, a small proportion only of the total volume, otherwise the compact would not retain its shape. Two different types of system can be distinguished: those in which the liquid phase is present throughout the sintering stage; and those in which the liquid phase is transient and disappears by diffusion into the solid either during the heating-up process or while the metal is being held at the ultimate sintering temperature. A good example of the former is hardmetal, in which the cobalt or other binder metal remains liquid even though it dissolves a small amount of tungsten carbide. To the second class belong the sintering of copper steels and of bronze. In these last two cases the fact that liquid phase sintering occurs is coincidental: the objective of the copper in iron and of the tin in bronze is to act as alloying elements to modify, i.e. improve, the end properties of the material. In other cases, the inclusion of a lower melting point metal (or metals) is wholly or largely a ruse to enable difficult materials to be sintered successfully to high density. An example of the latter is the production of so-called heavy alloy. This is based on tungsten which on its own is difficult to sinter to high density, but controlled additions of nickel and copper which become liquid at the sintering temperature enable a density closely approaching theoretical to be achieved. Although to a first approximation tungsten is insoluble in the nickel–copper phase, it is reasonable to infer that there must be some small degree of solution because, with progressively longer times at the sintering temperature, the particle size of the tungsten increases and this can be accounted for only by the transport of tungsten atoms through the liquid phase. Another industrial application of liquid phase sintering is the production of fully dense components in high speed steels. Water-atomized powder is compacted either in rigid dies or iso-statically and sintered in vacuum at a temperature just above the solidus temperature. This requires very careful control, but the result is worthwhile because the near approach to the final shape required drastically cuts down the amount of machining needed. High speed steels are, of course, used for metal cutting, and any necessary machining of the tool material itself has to be done by grinding—a time consuming and expensive process.

Activated sintering

This term is used to describe a process in which a purposeful, usually small, addition is made either to the powder mix or, more rarely, to the atmosphere in order to promote more rapid sintering or sintering at a lower temperature. The action of the activator may be to remove the surface oxide from the powder particles or to facilitate more rapid diffusion of the atoms of the metal. A typical example is the sintering of pure tungsten with the aid of a small amount—less than $\frac{1}{2}$%—of nickel. No liquid phase is involved, the effect occurs quite dramatically at temperatures well below the melting point of the nickel. Liquid phase sintering, however, can rightly be regarded as a special case of activated sintering.

Loose powder sintering

As the name implies, this means the sintering of uncompacted powder. We have seen that during the production of sponge iron powder the annealing treatment of the loose powder on a steel band results in the formation of a cake that has to be crushed to restore it to powder. This means that some sintering of the loose powder has occurred, fortunately only to a minor extent. However, the process is used intentionally in cases where a highly porous product is called for, for example filter elements, and also for steel-backed bearings. The latter process has been described earlier, and filters are dealt with in some detail in chapter 8.

Processes involving pressure as well as heat

There are several processes during which compaction and sintering occur simultaneously. Hot pressing, also called pressure sintering, has been mentioned already. The process normally involves heating the die as well as the powder or compact, and it is, therefore, slow and expensive. For these reasons it is not extensively used. A variant is *spark sintering*, where powder is compacted in a die and heated by passing an electric current via the opposing punches. Hot pressing, as is clear, applies to the production of individual pieces. Other processes are used increasingly for the production of wrought products from powder. These include the hot extrusion of powder billets which may but need not have been previously compacted, the Conform process, and hot iso-static pressing. The latter is used also in the production of near-net shapes of comparatively large size in special materials. When used for the production of wrought products, these processes provide either technical advantages over ingot-based material or, in some cases, economic advantage. They are the subject of chapter 7.

6 Post-sintering Operations

The various processing steps which may follow the sintering operation can usefully be divided into two categories:

(1) those that are peculiar to PM;
(2) those that are in general metallurgical use.

Operations Specific to PM Parts

Re-press and re-sinter

This requires no further elaboration. The object is to increase the density and thus improve the mechanical properties.

Hot re-press

This is more expensive—and more effective—than re-press and re-sinter, but the objective is the same.

Hot forge in a closed die

This is a special case of hot re-pressing in which, in addition to densification, a significant change of shape may be involved. This important process, called sinter-forging, is dealt with in chapter 7.

Sizing

This process, mis-called calibration in some European countries, involves cold re-pressing. A small amount of densification may take place but the primary purpose is not to improve the mechanical properties but to correct the dimensions of the part. Even with the best control that is feasible in practice, there will inevitably be some variation in the dimensions of the parts produced from a given powder in a given tool set. Among the causes of variation are: (1) variable density and (2) tool wear.

(1) *Variable density* of the die fill. One contributory factor is the uni-directional movement of the feed shoe. This is especially relevant with parts having thin sections at right angles to the pressing direction. Vibration also can influence the overall die fill.

(2) *Tool wear.* Methods used to minimize die wear have already been mentioned and include the selection of a powder with a low content of acid-insolubles, the incorporation of a lubricant, and the use of wear-resistant tool materials; but some wear is unavoidable and inevitably results in a gradual increase in the cross section of the component. Typically, it is possible for parts in the as-sintered condition to be accurate to a tolerance of 0.2% in the directions at right angles to the pressing direction, and 0.4% parallel to it. Sizing can greatly improve on this and also correct any warpage. It also serves to improve the surface finish assuming that properly polished tools are used. If no-growth mixes are used in the manufacture of the parts, sizing may be done in the dies that were used for compaction; but it is much more common to have separate facilities for sizing, not only tools but also separate presses, since the pressures used are normally much lower than are used for compaction.

Coining

This term is sometimes misused as a synonym for sizing. Its true meaning is to emboss a pattern on the end surface or surfaces of the part as in the manufacture from plain blanks of coins, from which process, of course, the name is derived.

Hot iso-static pressing (HIP)

More will be said on this subject later, but it should, for completeness' sake, be mentioned here. HIP is extensively used to eliminate flaws and micro-porosity in cemented carbides and in sensitive parts for aircraft engines. With the increase in the size of HIP equipment and the consequent lowering of unit cost, it is reasonable to predict that it will find use in the densification of PM structural parts. See the later section on the newer sinter-HIP process in chapter 7.

Steam treatment

This process involves exposing ferrous parts to super-heated steam at about 500 °C. It leads to the formation of a layer of magnetite (iron oxide) on all the accessible surfaces, i.e. including those of the pores that are open to the surface. A number of benefits result. Firstly, the density of the surface layers is increased and this confers increased compressive strength; secondly, the surface hardness is increased and with it the wear resistance; thirdly, the resistance to atmospheric corrosion is improved—see examples 15, 18, 19 and 25. Steam treatment is often followed by dipping in oil which enhances the blue–black appearance and still further increases the corrosion resistance. Steam treatment is peculiar to PM but the result is fundamentally the same as 'blueing', long used in the manufacture of, for example, tin tacks, but the latter is done at a lower temperature, ~200 °C, in air, and the oxide film is much thinner. The blue colour is caused by interference, whereas in steam treatment the colour is that of the magnetite.

Infiltration

Another method by which the strength of inherently porous sintered parts can be improved is to fill the surface-connected pores with another metal having a lower melting point than that of the original part. This is done by placing a quantity of the *infiltrant*, often compacted powder, on the top surface of the part and passing it through a sintering furnace at a temperature above the melting point of the infiltrant which melts and flows by capillary action into the pores. A requirement is, of course, that the infiltrant metal wets the metal to be treated. It is desirable that the infiltrant has a limited capacity to dissolve the metal being treated, otherwise the surface of the part may be eroded. The process is used quite extensively with ferrous parts using copper as infiltrant, but to avoid erosion an alloy containing, for example, iron and manganese is often used. Other proprietary infiltrant compositions are also on the market. If the molten copper is already saturated with iron, its ability to erode the surface is lost. However, as we saw in the section on dimensional change during sintering, the diffusion of copper into iron can lead to growth and infiltrated parts are generally less accurate as regards dimensions.

Infiltration is used also to make composite electrical contact materials, as will be described in a later chapter, and also in the manufacture of valve seats for internal combustion engines where a porous sintered part is infiltrated, for example, with lead.

Impregnation

This process is in certain respects analogous to infiltration except that the pores are filled with an organic as opposed to a metallic material. The outstanding example is oil-impregnated bearings, which are dealt with in detail later; but, increasingly, impregnation with thermo-setting or other plastic materials is being carried out. The benefits to be obtained include some increase in mechanical properties, sealing of the pores to provide pressure-tightness or to prevent the entry into the pores of a potentially corrosive electrolyte during a subsequent plating operation. Additionally, the machining properties of the parts are improved, a feature that is referred to in more detail later in this chapter. In a recently patented process, ferrous parts are impregnated with water glass—sodium silicate. The parts are then baked at a temperature sufficient to decompose the silicate and leave a residue of silica in the pores.

Processes in General Metallurgical Use

Most of the processes that are applied to ferrous components in general can be applied to sintered parts, but certain differences associated with the porosity should be noted.

Heat treatment—hardening and tempering

Many ferrous components are used in the as-sintered condition, but to make the most of their potential steels are increasingly supplied hardened and tempered in the same way as traditional wrought steels. The processes are basically the same except that since PM parts are porous, they should not be immersed in corrosive liquids because of the difficulty of removing such substances from the pores. This precludes, therefore, the use of salt baths for heating and of brine for quenching. In order to preserve the accurate dimensions, scaling must be avoided and, therefore, heating is done in a protective atmosphere and is followed by quenching, preferably in oil. Induction heating can also be used.

Porosity has an effect also on hardenability, that is on the depth of hardening. To achieve the full hardness of which a given steel is capable, it is necessary to cool at a rate greater than what is called the critical cooling rate, and this means that, no matter how efficient the quenching, full hardness is reached only to a limited depth because of the progressive reduction in the cooling rate from the surface to the interior. Porosity reduces the thermal conductivity of the steel and, therefore, the depth of hardening decreases as the porosity increases, as indicated in figure 18. (The term apparent hardness is explained in chapter 7.)

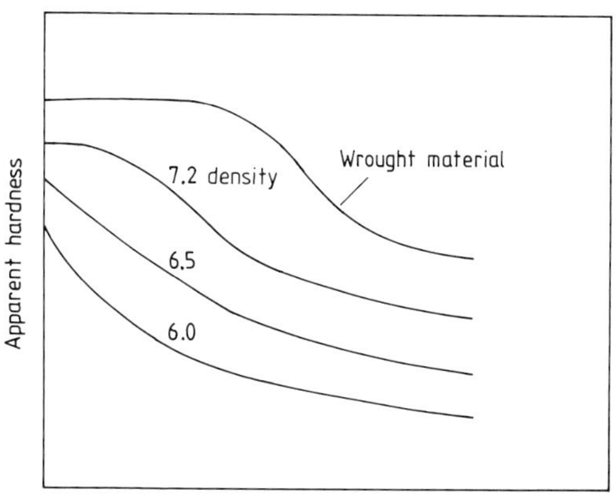

Figure 18 Hardenability as a function of density—end quench tests.

Surface hardening—case hardening

Carburizing, nitriding, and carbo-nitriding are all applied to PM parts and, again, because of the porosity, gaseous atmospheres are indicated. The porosity also affects the case depth. The gases can penetrate into the pores and so, for a given length of treatment, the case depth will be greater than with a dense steel of the same composition, and the line of demarcation between the case and the core is much less sharply defined (see figure 19). This feature is, in general, an advantage rather than the reverse.

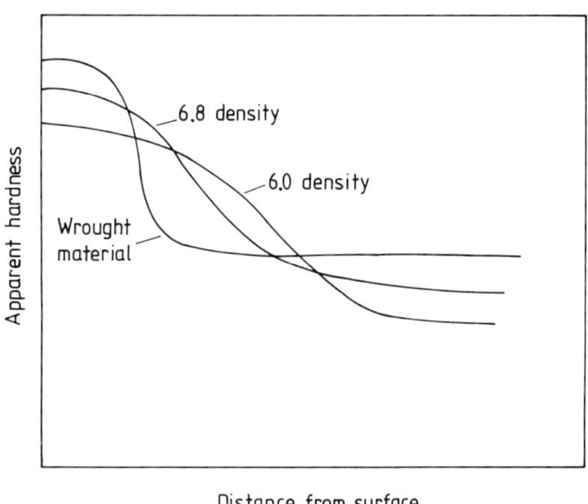

Figure 19 Effect of porosity on case hardening.

Plating

Sintered parts may be electroplated in much the same way as cast or wrought steels, and nickel, cadmium, zinc, and chromium are all applied. However, as has been mentioned earlier, it is advisable to seal the pores in some way in order to avoid ingress of electrolyte which it would be practically impossible to remove by normal washing. As electrolytes are usually corrosive to steel their presence in the pores is obviously undesirable. It has recently been reported that electroless nickel plating of non-impregnated steels may be feasible since the solution used, penetrating the pores, will plate their surfaces and not leave a corrosive residue when the parts are dried by gentle heating.

Other coatings

A large number of hardmetal and high speed steel cutting inserts are now given hard wear-resistant coatings which improve tool life enormously. In most cases a number of different layers are used—up to 10 has been reported—and their compositions include carbides, nitrides, carbo-nitrides and borides of metals, especially titanium and tantalum, and also oxides, especially aluminium oxide. The coatings, which are very thin, are normally formed by deposition from a vapour phase, and their application to sintered metals requires no special modifications.

Mechanical Operations

De-burring

Tumbling, sometimes in a liquid medium containing an abrasive powder, is used to smooth rough edges and remove any 'flash' resulting from compaction or from machining operations. Shot blasting may also be used. Burnishing and minor hardening of the surface may also take place.

Machining

Although a major attraction of powder metallurgy is that it produces parts so close to the required dimensions that machining is largely eliminated, as has been stated, there are certain limitations as to the geometry, imposed by the use of rigid dies. Therefore, it is not infrequently necessary to do a certain amount of machining, for example drilling of holes at angles to the pressing direction, cutting re-entrant grooves, making internal or external threads, etc. In some cases quite extensive machining is done as in example 9, where it can be seen that an addition of manganese sulphide is made to the powder mix to improve the machinability. This highlights a feature of PM steels in which they differ somewhat from wrought steels, that is in their machining characteristics. In general, the presence of porosity, which interrupts the cut, increases tool wear; and for that reason hardmetal tools are recommended as well as lower cutting speeds. Machinability is improved by certain additions to the steel; manganese sulphide has been mentioned, and lead in copper and brass are also used. As can be inferred, any process that fills the pores with a solid improves the situation; these include infiltration and resin impregnation. By taking the above considerations into account, all normal machining operations—turning, drilling, milling, broaching, threading, etc—can be done quite readily.

Joining

Most joining processes can be used with sintered parts, but the most common are perhaps copper brazing and projection welding. Copper brazing is often combined with infiltration. It is also possible to join two component parts at the compact stage by sintering them in close contact with each other. This is especially effective with assemblies with inner and outer components. By making the outer section of a composition that shrinks on sintering relative to the inner section, a mild pressure is exerted at the interface and a near-perfect bond results.

7 Compositions, Mechanical Properties, and Testing Procedures

Only ferrous materials will be dealt with in any detail, as they comprise the bulk of all sintered structural components. The properties of oil retaining bearings are referred to in the next chapter.

Compositions

The simplest composition is unalloyed iron. The term unalloyed rather than pure is used because there are always traces or more than traces of other elements present in the normal PM grades of iron powder. Unalloyed iron is the PM equivalent of mild steel, and is among the easiest materials to work with. It has limited strength and its properties are not susceptible to improvement by heat treatment, therefore it is seldom used for significant load bearing applications. It has the disadvantage also of shrinking during sintering. To provide better strength it is necessary to add alloying elements.

The cheapest alloying element is carbon. Figure 20 shows the strengthening effect of different percentage additions of carbon in what may be called plain carbon steels. Compositions with around 0.5% and 0.8% carbon are included in most PM standard specifications. The presence of carbon introduces the complication dealt with in chapter 5 on sintering, that a special atmosphere having the appropriate carbon equivalent must be used, so that neither carburization nor decarburization occur.

Copper, an alloying element rarely found in wrought steels and

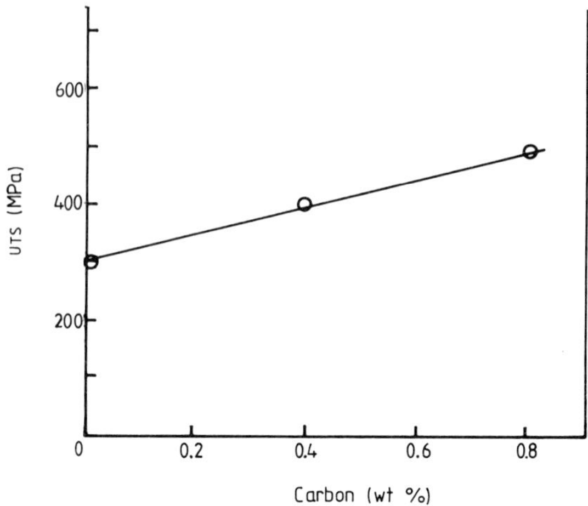

Figure 20 Strength as a function of carbon content (at a density of 7.0 g cm^{-3}) for as-sintered plain carbon steel.

then only for its effect on rusting, features extensively in PM compositions, and can be present in amounts up to as much as 20%, although less than 8% is most common. This level of copper is about the maximum percentage that will dissolve in iron, and it acts by solid solution strengthening. The use of copper affects the dimensional change on sintering as has been referred to earlier. With about 2% of copper little or no dimensional change occurs. A further advantage is that, as with plain iron, the sintering atmosphere need only be protective. The strengthening effect is, however, comparatively mild but much improved strength is obtained if carbon also is included. Copper bearing alloys are often referred to collectively as copper steels, but the carbon-free alloys are also referred to, and more logically, as copper irons, the term 'steel' being reserved for the heat treatable carbon containing alloys.

A second widely used metallic alloying element is nickel, either alone, with carbon, or more frequently with copper and carbon. Apart from its role in solid solution strengthening of the iron lattice, nickel counteracts the growth on sintering induced by high copper contents, and, importantly, improves the hardenability and the toughness. A third metallic strengthening addition is molybdenum, normally in conjunction with copper and nickel. Molybdenum has also a very beneficial effect on the hardenability of the steel. The reason for the choice of the above three metallic alloying elements, which are all relatively expensive, is that their oxides are reduced in quite mildly reducing atmospheres. By contrast, the oxides of the cheaper alloying elements, chromium, manganese, and silicon, that are extensively used in wrought low alloy steels, are not reduced in any of the economically feasible sintering atmospheres, and, indeed, any atmosphere that contains oxygen even in combined form as, for example, water vapour or oxide(s) of carbon, will oxidize these elements. To a first approximation then we may say that compositions containing these elements in elemental form cannot be used, and it is common practice to add the required alloying elements in elemental form in order to preserve the good compressibility of the iron powder, as has been explained; but there are two consequences. Firstly, the added elements do not diffuse into the iron during the normal sintering cycle at a sufficient rate to give a homogeneous alloy. Whether this is necessarily disadvantageous has been debated—some workers go so far as to argue that the micro-heterogeneity of such alloys can be beneficial as far as mechanical properties are concerned. We need not pursue that discussion, sufficient be it to say that these sintered alloy steels have properties that are adequate or more than adequate for the purposes for which they are used. On the second consequence of adding the elements in elemental form there is no room for debate. Mixes of markedly different powders are prone to segregate during handling and the resultant macro-heterogeneity is highly undesirable. An ingenious solution to this problem, the production of partially pre-alloyed powders, was mentioned in chapter 3. It should be emphasized, however, that straightforward mixes of elemental powders give generally satisfactory results; the additional steps referred to are designed to give still greater uniformity to the finished product, especially as regards dimensions.

The above picture regarding alloying elements has changed somewhat in the last decade or so, when it was discovered that if chromium, manganese, and also vanadium are introduced into the powder mix in the form of carbides, they are not oxidized during sintering in traditional atmospheres. They are added as mixed carbides—manganese/chromium/molybdenum carbide or manganese/vanadium/molybdenum carbide—and in order to promote rapid diffusion into the iron, they are added as very fine powder—5–10 μm—made by mechanical comminution. Their use has not yet made a very great impact, perhaps because they need a sintering temperature well in excess of 1200 °C compared with the 1120–1130 °C for traditional mixtures, and this temperature is beyond the capability of mesh belt conveyor furnaces. Many parts producers do not have the walking beam furnaces that need to be used. The use of higher sintering temperatures puts up the cost of sintering, and this may be a further deterrent.

Another development, and one that comes as a surprise to steel metallurgists, is the use of phosphorus as an alloying element. In ingot-based materials, traditional lore will tell you that phosphorus is a bad thing, and steel specifications in general put a maximum on the content permitted as an unavoidable impurity. The reason for this is that phosphorus forms an iron/phosphorus eutectic which has a lower melting point than the bulk of the steel and, therefore, segregates towards the centre of the ingot with deleterious effect on the mechanical properties resulting from the brittleness of the iron phosphide. In PM, the phosphorus is added as fine ferro-phosphorus and is uniformly distributed. When the temperature reaches about 1050 °C a liquid phase appears and this greatly accelerates diffusion and sintering. A consequence of this is that shrinkage is increased, especially if the phosphorus content exceeds 0.25%, but this can be compensated for by an addition of copper. It can be inferred that normal sintering temperatures around 1120 °C are sufficient.

In both the above developments it should be said that die wear may be increased, as carbides and iron phosphide are very hard and, therefore, potentially abrasive substances.

In the above discussion of alloy compositions we have tacitly assumed parts sintered to equal density, but we know already that density itself is an important factor in determining mechanical properties; this is brought out strikingly in figures 21 and 22 and table 2. Notice that unlike wrought materials, the ductility of which is often reduced by any steps that increase the strength, with sintered steels the ductility also increases as the density and strength increase. This stems from the fact that the low ductility is basically a consequence of the porosity, so that as the porosity is reduced the ductility improves. It should be emphasized that because PM parts are made directly to finished shape, ductility does not have the same importance as in wrought steels—the metal is not required to be plastically deformed as is wrought material. For severe service conditions, impact and fatigue strengths are of far greater consequence and both are seriously reduced by porosity (see figure 23). Therefore, in order to enlarge the area of use of sintered steels to include more demanding applications, means of achieving higher densities have been extensively investigated.

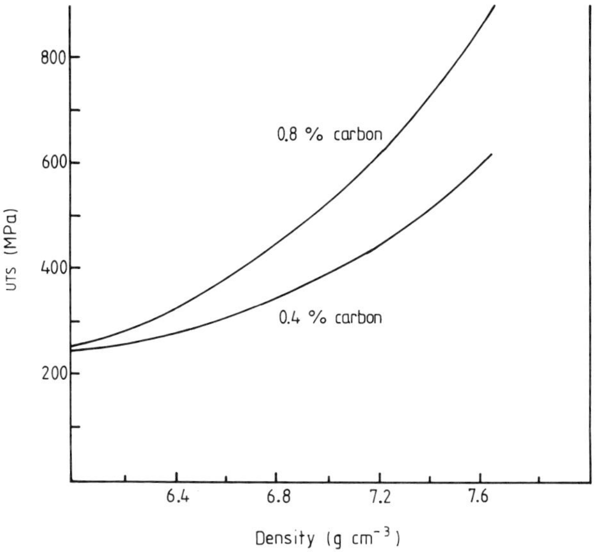

Figure 21 Tensile strength as a function of density.

Reaching Higher Density

The factors that limit the achievement of high density compacts, e.g. the presence of a pressing lubricant, have been discussed already, as have the means by which the final density can be increased by re-pressing and sintering and by hot re-pressing, but these processes do not easily give properties equivalent to those of comparable wrought steels. However, two fundamentally

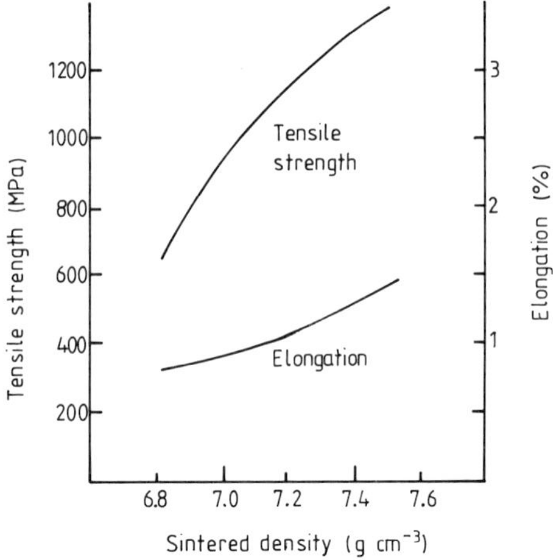

Figure 22 Tensile properties of a sintered alloy steel as a function of density.

Table 2 Typical mechanical properties of some iron-based PM materials. The figures in brackets are for the steels in the heat-treated condition; the remainder refer to the as-sintered material.

Composition (wt %)				Density	Yield strength	UTS	Elongation
C	Cu	Ni	Mo	($g\,cm^{-3}$)	($N\,mm^{-2}$)	($N\,mm^{-2}$)	(%)
0.25	—	—	—	6.1	90	140	3
				6.9	140	210	9
0.25–0.6	—	—	—	6.1	130 (380)	170 (400)	1.5 (0.5)
				6.9	190 (520)	290 (560)	3.5 (0.5)
0.25	1–3	—	—	6.1	115	160	2.5
				6.85	160	260	5
0.25–0.6	1–3	—	—	6.1	230	275	1
				6.85	310 (650)	420 (690)	3 (0.5)
0.25–0.6	2–4	0.5–2	—	6.1	100	200	0.5
				6.95	210 (600)	340 (750)	2.5 (1)
0.25–0.6	1.5	4	0.5	7.1	350 (770)	500 (980)	5 (1)

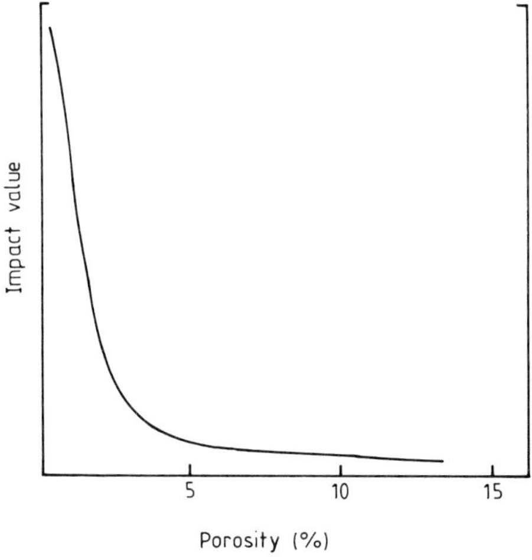

Figure 23 Impact value as a function of porosity (schematic).

different processes have been developed that go a long way towards achieving that goal.

Sinter forging/powder forging

The production of powder-based blanks or pre-forms for subsequent hot forging was suggested quite early in the development of PM structural parts, but serious attempts to put the idea into industrial practice did not begin until the 1970s when a pilot plant to produce a differential pinion gear was successfully operated. A great flurry of activity followed and practical experience on an industrial scale in a considerable number of companies, in both the PM and the forging sectors, confirmed that the process was technically sound. However, the close control that was necessary and problems of die life raised doubts about the process's economic viability, which led to the closure of several plants and a period of disillusion. In Europe only one plant remains operative. In the USA and Japan, however, rehabilitation has occurred, and several companies are in regular and, presumably, profitable production of a considerable range of components. Figure 24 illustrates a typical production sequence for sinter forging.

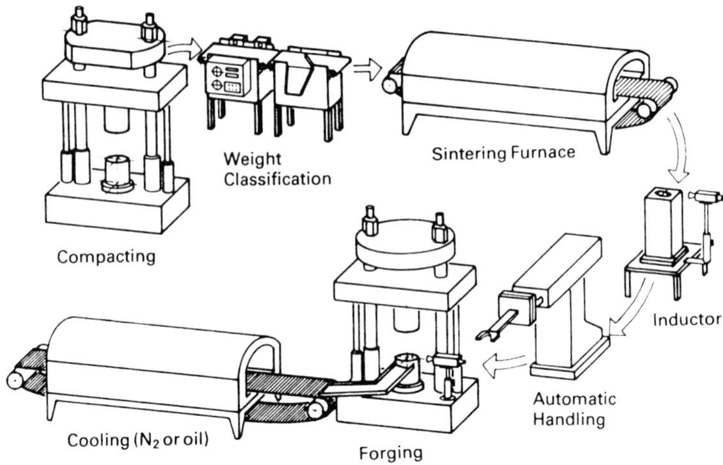

Figure 24 The production sequence for powder forged precision components (induction heating to forging temperature). *SMK.*

In essence, sinter forging is the same as the closed-die forging (precision forging) of wrought steels but it differs in two significant ways. Whereas the solid blanks for die forging are normally circular discs or cylinders cropped from bar, the blanks for sinter forging may have a shape much nearer to that of the finished part. A second and very important difference is that it is possible, by accurate metering of the powder into the compacting die, to make blanks of a much more constant weight than those produced by cropping steel bar. The end result is that the forging operation does not produce any 'flash' and the dimensions are, therefore, much more consistent. To increase this uniformity still further, some plants have installed weighing devices to the compacting press and any compact that does not fall within a pre-determined and very narrow weight range is automatically rejected.

Pre-form design

In the early stages of the development of the process the optimum shape of the pre-form was determined by trial and error, but with increased understanding of the deformation processes involved, guide lines to the design of pre-forms have evolved which, at any rate, reduce the amount of development work necessary for the production of a new part. It is now possible to design the pre-form and the forging tools so as to control the metal flow and to ensure that the mechanical properties are at their optimum in the regions of maximum stress, e.g. at the root of the teeth of gear wheels.

Sintering and forging

After compacting there are alternative procedures. In one, the compacts are sintered in the normal way in continuous furnaces, and subsequently re-heated, often by HF induction, in preparation for the forging operation. The alternative is to take the sintered blanks directly from the hot zone of the sintering furnace, which, of course, will be without a cooling zone, straight to the forging press. It is this difference that led to the different nomenclatures—sinter forging and powder forging—but in effect there is no difference; in both cases it is a sintered blank that is forged. The two terms must, therefore, be regarded as synonymous.

At the forging temperature, rapid surface oxidation of steel occurs on exposure to air, and with a porous blank this is potentially more damaging than with a solid one since oxidation can penetrate below the surface. Consequently, for successful operation, speed of transfer of the blank from the heating furnace, with its protective atmosphere, to the forging die is essential. Exposure times have been cut to the order of one second, and it is believed that this time is short enough to ensure protection of the pores, at least, by the gas entrapped within them.

The powders used for sinter forging are always pre-alloyed alloy steels, usually containing nickel and molybdenum, only graphite being added in elemental form. Copper has been included in some cases to improve the hardenability of the steel during subsequent heat treatment. Sintering temperatures are relatively high, being of the order of 1225 °C.

A comparison of sinter forged components with similar components made by precision forging of solid metal shows the following features.

(1) Mechanical properties are generally much the same, but those of sinter forgings are more isotropic; the marked difference in, e.g., ductility and impact value that exists in rolled bar between the longitudinal and transverse directions being almost entirely

absent so that, in the direction of maximum stress, it has been demonstrated that a sinter forging may actually be superior.

(2) The consistency and accuracy of dimensions of sinter forgings are better.

(3) Sinter forgings have generally better surface finish.

The conclusion must be, therefore, that, apart from considerations of cost, a well made sinter forging is certainly as good as and may be better than a traditional precision-forged component.

Iso-static pressing

Cold iso-static pressing

The mechanics of cold iso-static pressing, usually shortened to CIP, were touched on in chapter 4. The major merit of the process is its ability to produce shapes that are impossible in rigid dies, e.g. parts with re-entrants and long parts such as tubes, but, additionally, the density of the compacts can be much greater and may closely approach 100%.

There are two variants of the process referred to somewhat illogically as wet-bag and dry-bag CIP. In the former, the mould is closed with a bung of the same material and totally immersed in the pressurizing liquid. In dry-bag compaction, the top of the mould is rigidly attached to the cover plate of the pressure vessel, thus making it possible to achieve a much higher rate of production. There is a disadvantage in that the neck of the mould is not under truly iso-static pressure and the top section of the compact may be less dense than the bulk. A very successful and well established application of dry-bag CIP is the production of spark plug insulators, but these are made of alumina, not metal. Another application that has reached an advanced stage of development is the production on an automated line of iron cylinder liners for internal combustion engines.

Hot iso-static pressing

HIP has now passed into the international language of powder metallurgy, both as a noun and a verb. The term means, of course, hot iso-static pressing, and it stands in the same relation to cold iso-static pressing as does hot pressing to cold pressing, but there the similarity ends. HIPping is carried out for the most part at high temperatures, such that the use of a liquid pressurizing medium is not feasible, and in practice an inert gas, namely argon, is used. The reason for using this truly inert gas, expensive though it is, is that any other gas at high temperature and pressure would react with the material being compacted and with the materials of which the equipment is constructed.

The equipment consists basically of a pressure vessel, inside which is located an electrically heated furnace surrounding the central space in which the work load is placed. Insulation between the furnace and the pressure vessel is provided and thermocouples are located at strategic points. At its simplest the procedure is to load the work into the chamber, close the top opening, and introduce compressed argon. When the furnace is switched on the temperature rise automatically still further increases the pressure and much of the equipment now commercially available can operate at up to 2000 °C and 2000 bars pressure. Metals and other materials become softer as the temperature rises, and complete densification can readily be achieved. However, because of the complete non-reactivity of the gas, any argon contained in surface-connected pores of the workpiece, or in the interstices between the particles if a powder is being HIPped, would merely be compressed and would remain as a discontinuity in the metal to the detriment of the mechanical properties of the finished article. For this reason, already sintered parts must have negligible interconnected porosity open to the surface if they are to develop their optimum properties. In practice this means that they must have a density in excess of 90% of the theoretical, in which case the pores will either contain air, or if the part has been sintered in vacuum, nothing. Air will, at the high temperature and pressure involved, either react with or dissolve in the metal, allowing complete pore closure to occur. HIPping is now routinely applied not only to PM materials but also to castings, in order to achieve full density and consequently the best possible mechanical properties. Hardmetals (cemented carbides) are notable beneficiaries as are superalloy blades for high temperature gas turbines.

Recently, a development known as *sinter-HIP* has begun to make its mark. In this process, initial sintering of the compacts and their subsequent densification are done in the same vessel in a single cycle. To avoid the complication of gas in the pores, the sintering is done in vacuum and, the necessary 90% density having been reached, pressurization with argon completes the process. The economic advantage of this process compared with the normal sintering followed by HIPping in a separate operation will be obvious, and it is accepted in the hardmetal industry that sinter-HIPping will supersede the older process.

Testing of Sintered Parts

The properties of sintered parts that are of interest to the engineer are, for the most part, the same as for wrought metal components, and in many respects the tests that are used to determine their suitability for the intended service, or to ensure compliance with an agreed specification, are identical. However, there are certain differences stemming from two features of PM parts: (1) they are porous; (2) they are small. We will look first at the second.

With wrought components it is frequently possible to test the actual piece of metal of which the part is made. In the case of a substantial forging sufficient surplus metal can be included to allow test pieces to be cut from the forging itself; if it is a small part machined from bar or the like, test pieces can be cut from the bar. Neither of these procedures is relevant to sintered parts and, therefore, the nearest we can get to a representative sample is to press test bars from the same powder mix to the same density as the component, and to sinter them at the same time, i.e. among the actual parts.

Tensile tests

Figure 25 shows the contour of the test piece prescribed in the ISO Standard 2740. The thickness is between 5.4 and 6 mm and the gauge length is 25 mm, scribed symmetrically.

Hardness testing

This is less straightforward because special problems arise when testing is done by the normal indentation techniques, in consequence of the porosity. Nevertheless, as in the case of dense

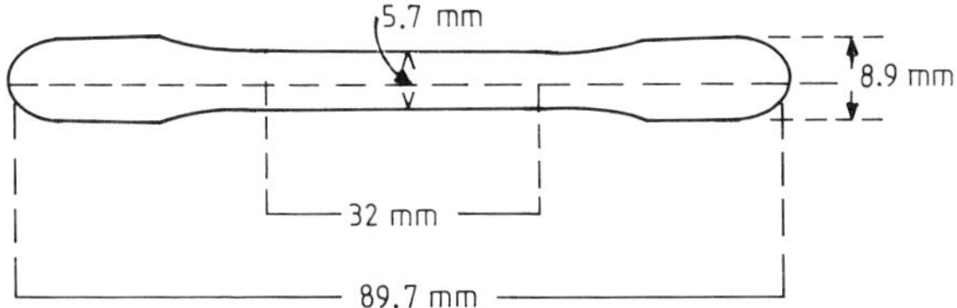

Figure 25 Standard tensile test piece—ISO 2740. Thickness after sintering 5.4–6.0 mm.

materials, hardness is a useful indication of certain mechanical properties, and because it is an easy and inexpensive test to do, and can be done on actual parts, it is in general use. With porous metals, however, we need to distinguish between what is called *true hardness*, i.e. the hardness of the same metal in the fully dense state, and *apparent hardness*, i.e. the hardness indicated by a macro-hardness test. If the volume of metal displaced by the indentor includes pores, the indentor will penetrate further than it would in dense metal because the holes offer no resistance to deformation. Clearly, therefore, the hardness value indicated will be lower than the true hardness. The apparent hardness bears a relation to ultimate tensile strength similar to that for dense metal because, of course, the holes are equally effective in reducing resistance to deformation in both cases. True hardness is more relevant to wear resistance.

Apparent hardness may be determined by the Brinell, Vickers, or Rockwell methods and more consistent results are obtained if a reasonably high load is used. If the Brinell or Vickers test impression does not have clearly defined edges, that impression is ignored. Five valid impressions are measured and the lowest value of hardness is discarded before taking the average.

True hardness has to be determined by a micro-hardness test which uses a relatively very small load and, consequently, makes a very small impression. The surface is required to be very much smoother than is needed for testing the apparent hardness, and a polish equivalent to that used for microscopical examination is normal. Care must be taken in the preparation so as to avoid significant work hardening of the surface. Since we are trying to determine the hardness of the metal without pores we have to select, under the microscope, an area such that the impression is at least as far from a pore as its own diagonal. Even then we have no means of knowing what is underneath the surface at the point of indentation, and so a large number of tests may be required, obviously low readings being discarded. As regards the load used, there is a conflict of interests. The smaller the load the more chance there will be of finding a spot free from the complicating effect of porosity, but at the same time, the indentation will be smaller and more difficult to measure accurately. A compromise is normally accepted and a load of 100 g used.

Impact testing

As the pores behave substantially as notches, impact values on normal sintered materials are very low and there is considerable scatter. In consequence the test is seldom specified. Nevertheless, for research and development purposes if nothing else, a test piece has been specified. It is a square section $10 \times 10 \times 55$ mm, normally un-notched, and broken on a Charpy machine.

Fatigue tests

Fatigue testing is a long-term and expensive procedure, and is seldom if ever included in specifications for PM parts. However, it is of interest in the process of development of materials and processing parameters. As with impact tests, the effect of the porosity is to give relatively low values for fatigue strength.

Bend test

This measures what is called the transverse rupture strength. ISO standard 3325 specifies a rectangular test piece $30 \times 12 \times 6$ mm to be broken in three-point bending. The load is gradually increased until fracture occurs—indicated by a sudden reduction in the load—and the transverse rupture strength is given by the formula

$$\frac{3PL_2}{2ba}$$

where P is the breaking load, L the distance between the supports (25 cm), b the width of the specimen (~12 mm), and a the thickness (~6 mm). The test may be specified on sintered materials, but it is more commonly used in PM on compacts, as a measure of green strength.

Surface roughness

This is another property that has relevance in some circumstances, but with sintered materials, determination and even definition is very difficult. The normal method using a profilometer is of little value—the pores in the path of the measuring device will appear as roughness even though the surface between the pores has a mirror finish. A great deal of work has been done on the subject but, so far, no acceptable procedure for roughness testing has emerged.

Density

This property is specific to PM parts and is, of course, in inverse relationship to porosity. The standard method for determining density is by Archimedes' principle, i.e. comparing the weight of the object in air with what it is when immersed in water. This is used also for PM parts, but since it is the overall density of the part including the pores that it is intended to measure, steps must be taken to prevent water from entering the pores. For this purpose it is usually sufficient to dip the part in paraffin or another hydrophobic liquid and to remove the surplus from the surface using a non-absorbent material. A useful indication of the efficiency of the surface impregnation is provided by observing whether any increase in weight occurs during the weighing in water; if it does, it would indicate that water is entering the pores.

Other tests

It is not uncommon for other specific tests, often breaking tests on actual parts, to be agreed between supplier and user.

8 Porous PM Components

We have seen that most sintered components are porous, but in this chapter we will refer to products that are porous by design. There are two such products that are of major importance, namely oil retaining bearings and filter elements.

Porous Bearings

The basic features of porous bearings were mentioned in the introduction. The majority are made of 90/10 tin/bronze. The choice of this composition was dictated by its established position as an excellent bearing material when running against steel. From the users' point of view, two properties are of importance: (1) the amount of oil, which determines the useful life of the bearing, and (2) the strength. It will be obvious that these two features are in opposition. For a given material, the strength is higher the lower the porosity, but the oil content is higher the greater the porosity, assuming in both cases that the bearing has the same dimensions. Table 3 gives typical figures for 90/10 bronze.

Table 3

Density ($g\,cm^{-3}$)	Porosity (%)	Tensile strength ($MN\,m^{-2}$)
5.8	31	55
6.0	29	62
6.3	25	76

The oil content is not necessarily exactly equal to the porosity, but it is very closely related to it.

Originally, the starting materials for bronze bearings were pure electrolytic copper and tin powder, normally produced by air atomization, but as we saw in chapter 2, there has been a move away from electrolytic copper in favour of water-atomized powder. However, the behaviour of elemental mixtures of copper and tin is much the same, no matter what the source of the copper powder.

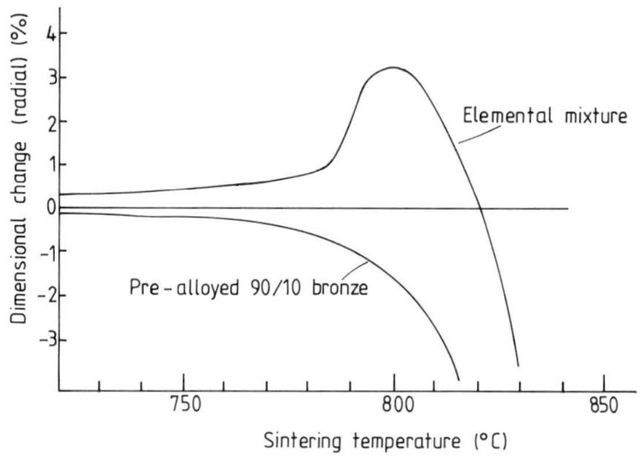

Figure 26 The relation between final dimensions and sintering temperature for bronze bearings.

The sintering of copper/tin mixes has been extensively investigated, especially as regards dimensional change. The tests have generally been done using plain cylindrical bearings (bushes, as they are called). As was mentioned earlier in chapter 5 on sintering, such mixtures show an increase in dimensions when sintered in the normal way. The dimension of most relevance is the radius, and figure 26 shows the changes that take place, the measurements being taken after the specimens had been cooled from the sintering temperature to ambient. It can be seen that some growth takes place at relatively low temperatures, but at around 800 °C a sudden growth occurs. This is attributed to the peritectic reaction at 798 °C in which a tin-rich liquid phase is formed. As the temperature rises still further, the expected shrinkage sets in. Varying the sintering temperature is sometimes used as a means of controlling the final dimensions, but there are limits to which this can be done.

Lowering the sintering temperature reduces the strength, while the risk of surface roughening and distortion restricts the freedom to increase the temperature. Temperature is not the only factor affecting the dimensional change. The particle size of the powders has a very significant effect, growth being lower if finer powders are used. This too can be used as a growth controlling mechanism, but, again, there are limits: if the powders are fine, the flow is adversely affected. However, manufacturers of pre-mixed bronze powders regularly use the fineness of the tin powder for regulating dimensional change. Sintering time also has an important effect: the longer the furnace time the lower the growth. The actual positions of the curves in figure 26 apply only to a particular sintering regime; slower sintering would reduce the peak on the curve for elemental powder mixes and increase the shrinkage with the pre-alloyed material. This feature also can be used to control dimensions, but manufacturers generally, and understandably, prefer to standardize their process. In any case, slowing down the furnace belt would reduce the rate of production and, in the highly competitive field of bearings, no manufacturer would contemplate that with equanimity. As explained in chapter 5 on sintering, there are good technical reasons that limit the extent to which the process can safely be speeded up.

Yet another factor affecting dimensional change is compact density. As can be deduced, the lower the compact density the greater the scope for shrinkage so that, in any given set of sintering conditions, a less dense compact will shrink more, or grow less, than a denser one. The extent of the growth is increased by the presence of graphite which is favoured by some producers as an addition, in amounts up to about 2%, designed to provide further lubrication. Bearings with as much as 5 or even 6% of graphite are sometimes specified in applications where 'quiet' running is an important requirement, e.g. sound reproducing equipment.

The ability of these elemental mixes to provide large increases in dimensions has been used by some manufacturers to produce bearings with a very large percentage of porosity, while at the same time using compacts of higher density to ensure adequate green strength. However, with these large amounts of growth, control of the final dimensions is less reliable and, with the increasing emphasis that users, especially in the motor industries, are now putting on dimensional accuracy, the use of high growth mixes is declining. A well known method of reducing growth is to include in the mix a significant percentage of pre-alloyed bronze. As the percentage increases, the growth decreases until, at a certain point, shrinkage begins to occur. Thus it is possible to provide mixes that show no dimensional change on sintering—so-called no-growth mixes—and with such mixes it is easier to make bearings with consistent dimensions.

Another important aspect is strength and, as mentioned above, strength and the other requirement, high oil content, are in opposition. Strength is somewhat increased by cold working during sizing but heavy deformation closes the porosity. Better strength can readily be provided by using greater wall thickness, but this increases the cost, and in fact the tendency over the last few decades has been to reduce the thickness, especially at times when the cost of the metals, especially the tin, has escalated. For less critical bearing conditions, mixtures of bronze and iron and, to some extent, copper and iron or even plain iron are used, and these give higher strengths than straight bronze.

As the term implies, the object of sizing is to give the bearing the exact dimensions required. The greater the deformation that takes place on forcing the bearing through the sizing tools the less accurate will be the final dimensions because of spring back; hence, the modern preference for low-growth or even no-growth mixes. When discussing dimensions, we are normally concerned with diameter; in most cases the length of the bearing is much less critical.

The final step is the introduction of the oil. This is done by loading the parts in a wire or perforated sheet basket, which is then immersed in the lubricating oil contained in a vessel with a vacuum-tight top closure. Vacuum is then applied and the gas in the pores is gradually sucked out to be replaced by oil when the vacuum is released. The basket is then withdrawn and allowed to drain. It is important that the oiled bearings be kept out of contact with oil absorbing materials. If, for example, the bearings are stood on blotting paper a lot of the oil will be drawn out. Most specifications call for a minimum oil content of 90% of the surface-connected porosity, and it is not necessary to use a high vacuum to achieve or even exceed this figure. Oil contents range from about 15 to 30% by volume of the total volume of the bearing. Figure 27 shows a collection of such bearings.

The largest area of use for self-lubricating bearings is in small

Figure 27 A selection of oil retaining bearings. *Ringsdorff-Werke.*

electric motors which are used in a multiplicity of machinery for business and domestic use, as well as in motor cars.

Filters

Filters are used, in the main, to remove solid particles from liquids or gases. Many different types are commercially available, ranging from the filter papers used in the chemical laboratory and in certain types of coffee making machines, cloth, porous glass, woven wire gauze, to perforated metal sheet. For many industrial applications the first two are ruled out because they are not resistant to corrosive substances, but additionally they do not readily permit a large flow of liquid. Glass is very brittle and difficult to incorporate securely without risk of breakage. Wire gauze and perforated metal sheet suffer from neither of these disadvantages provided a metal that is resistant to the fluid to be filtered is chosen, but they offer only a single barrier to the passage of solid matter. Sintered metal filter elements, as they are called, provide in-depth filtration, and for many purposes are preferred.

The most widely used materials for filters are bronze and stainless steel, but titanium is also used to a small extent for the filtration of very corrosive liquids. Sheets of porous nickel are used in the separation of the isotopes of uranium, but that is not strictly filtration, since the enrichment that occurs at each stage is very modest and depends on the different rates of diffusion of the two isotopes through the porous barrier, rather than the retention of one and the passage of the other.

Bronze filters

These are made from spherical bronze particles usually produced by air atomization. Best results are obtained when all the particles are of the same diameter, but as this is impracticable, a restricted range of sizes is used. Commonly specified is that the smallest sphere shall be not less than two-thirds of the diameter of the largest. An alternative method of making bronze filters that is favoured in the USA is to chop more or less equiaxed pieces of copper wire, partially round them by tumbling or other mechanical process, and then coat the particles with tin, which diffuses into the copper to form bronze on sintering. To make the filter element, loose powder sintering is used, i.e. the particles are loaded in a suitably shaped mould made of graphite or stainless steel and sintered at a temperature of about 800 °C, i.e. near to the solidus temperature. A range of shapes is made, varying from discs to tubes, the latter often having one end closed and the other end flanged—see figure 28. A micrograph of sintered spherical bronze appears in chapter 5 on sintering (figure 16). By sintering together two different cuts of powder disposed in separate layers, a filter may be made such that the fluid to be filtered encounters first a more porous layer, which removes coarse solid particles, and then a layer with smaller passages to remove finer solids. This procedure increases the time that the filter can be used before it becomes blocked and needs to be cleaned by back-washing.

Stainless steel, titanium and nickel

The production of spherical powders of these metals is not practised, filter elements being made from irregular powders. For the most part they are made in the form of sheet by one of two methods.

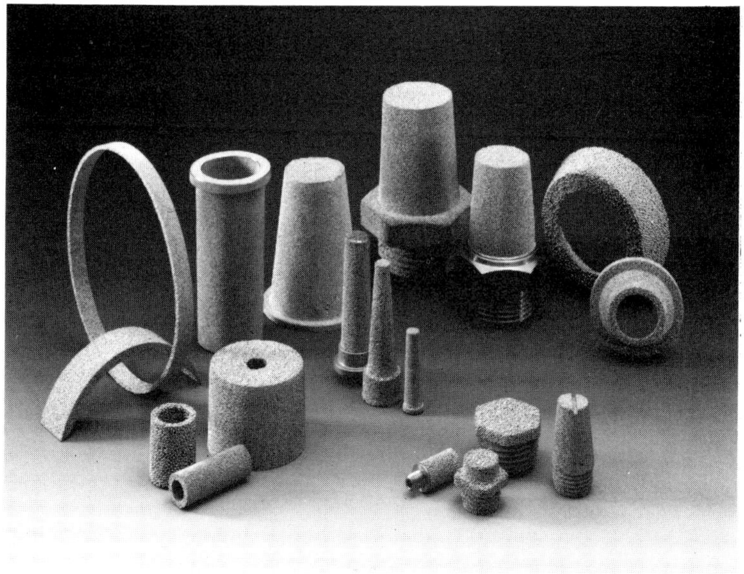

Figure 28 Sintered bronze filters. *Norddeutsche Affinerie*.

(a) The powder is mixed with a binder and spread on a flat surface on which it is sintered, the binder being first burnt off. The sintered sheet may then be pressed and re-sintered to give it additional strength.

(b) A layer of powder is directly compressed and sintered. The use of powder rolling has also been reported.

If a cylindrical element is required, the sintered sheet is formed into a cylinder of the required diameter and a longitudinal weld made.

Another process that has been used is the cold iso-static pressing of the powder in the annular space between a solid core and an elastomeric outer mould.

Other uses

Porous sintered metals find several uses other than in filtration:

(a) as a means of supplying a stream of gas bubbles in a liquid;
(b) as silencers to damp the throb of gas compressors;

(c) as electrodes in alkaline batteries;
(d) to separate oil from water.

They were at one time fitted to the leading edges of the wings of aircraft to provide a continuous supply of de-icing liquid.

Testing of filters

Several standardized tests are available for filters. In one of these (ISO 4003) the filter is impregnated with a liquid and immersed in it. Air at gradually increasing pressure is applied to one side of the filter and the pressure at which bubbles begin to emerge is noted. A figure representing the maximum pore size is calculated by the formula

$$d = \frac{4y}{p}$$

where y is the surface tension of the test liquid, and p the pressure difference.

Another test directly determines the permeability to the test fluid; the volumetric flow rate for a given pressure drop being measured (ISO standard 4022).

Surgical Implants

Yet another use for porous PM materials, not strictly relevant to engineering, is as surgical implants such as, for example, hip joints. These are made of corrosion-resistant metals that are compatible with living tissue, and it is found that if they are made by PM and are porous, bone tissue will actually grow into the pores of the implant, providing a strong joint and obviating the need for cement. The whole implant may be made by PM or alternatively a porous layer may be sintered on the surface of a solid, e.g. a cast part. The metals tantalum and niobium have both been found suitable.

9 Special Products and Other Compositions

Magnetic Materials

Soft magnets

The justification for making soft magnets by PM is basically the same as for structural parts, principally the fact that parts can be pressed and sintered to the required shape and size, obviating the necessity for machining. This benefit is perhaps even greater with magnetic materials because they are normally very soft and not easy to machine cleanly.

The commonest compositions are:

pure iron
iron with an addition of silicon, for example 3%
iron with about 5% of phosphorus
a 50/50 nickel/iron alloy.

Typical soft magnetic parts are shown in figure 29.

The best magnetic properties are obtained when the impurity level is kept low, and for that reason electrolytic iron powder was used initially. This powder is, however, quite expensive and, with the best technique now available, atomized powder low in the interstitial impurities carbon, oxygen, and nitrogen can be used with only a small reduction in performance. Silicon is added as finely crushed ferro-silicon, and phosphorus as ferro-phosphorus. Nickel iron is in the form of fully pre-alloyed atomized powder. The powders are compacted in rigid dies and sintered in the same way as in the production of structural components. The magnetic properties, as could be inferred, are better the higher the density, and for that reason, sintering temperatures significantly higher than those used for structural parts are used, e.g. around 1300 °C. Low

Figure 29 Parts for soft magnetic applications. *Höganäs.*

dew point cracked ammonia or vacuum are often employed and complete homogeneity of composition is aimed at. In the case of iron phosphorus, densification is assisted by the transient liquid phase that is produced.

Permanent magnets

Permanent magnets represent an area of technology that has, in recent times, developed faster and further than most others in the materials field, and the usage of magnets also has grown in spectacular fashion. It has been stated that in the USA every household has, on average, about 80 magnets. Powder techniques are in the forefront of many of the developments.

The earliest recorded permanent magnet material is lodestone, a naturally occurring mineral of variable composition but basically a double oxide of iron—FeO,Fe_2O_3—or, as we would now call it, ferrous ferrite. Lodestone was known more than 3000 years ago. During the 18th century a synthetic magnetizable iron oxide was produced in England from iron filings, and this material, mixed with linseed oil, moulded to shape and baked, provided what were described as 'very strong magnets'. Interestingly, a comparable process, the incorporation of strongly magnetic particles in an organic matrix, has come to the fore during the last two decades.

Before this, however, a very significant development, dating from 1933, was the production of the so-called Alnico magnets, based on iron alloyed with aluminium, nickel, and cobalt. There are several grades, some also containing copper. These alloys were first made by casting and, for the most part, still are; but they are hard and very brittle and so require to be ground to size. PM is an attractive alternative, and millions of sintered Alnico magnets are made annually, mostly in the smaller sizes. The ingredients are ball milled together, and the mixture compacted and sintered in dry hydrogen at a temperature of 1300 °C or higher.

Patient research in the 1940s led to renewed interest in ferrites (mixed oxides, one of which is of iron) and barium ferrite and later strontium ferrite quickly took a large share of the market, over 50% at one estimate. These are made by PM techniques, and are much cheaper than Alnico.

The most recently developed magnet materials having an energy product, BH_{max}, far higher than any other material are generally referred to as rare-earth–cobalt magnets. The rare earths (RES) are the oxides of a closely related group of metallic elements, and it is in the metallic form that they are used; so strictly speaking, we should call them rare-earth metal–cobalt magnets. These RE elements always occur together in nature, and because they are chemically very similar, they are not easy to separate and were for many years offered as a mixture, generally known by the German name *Mischmetall*, the predominant element being cerium. Methods for their separation on a commercial scale are now available.

The manufacturing process commonly used for the magnets is firstly to reduce the rare earth mixed with either cobalt powder or its oxide with calcium:

$$\mathrm{RE_2O_3 + 10Co + 3Ca \rightleftharpoons 2RECo_5 + 3CaO.}$$

Excess calcium and the calcium oxide are leached out, leaving the RE–cobalt mixture as powder. This is then compacted and the particles aligned in a magnetic field, followed by sintering and heat treatment. Alternatively, the powder mixed with a thermo-setting resin is moulded to shape in a magnetic field and then warmed to cure the resin.

A milestone along the road was reached with the samarium–cobalt compositions $SmCo_5$ and Sm_2Co_{17}. These are intermetallic compounds that display remarkable magnetic anisotropy and, when suitably aligned, give the highest energy product per unit mass of any known permanent magnet material—some five times that of Alnico. A still higher BH_{max} has been achieved with alloys containing neodymium, iron, and boron, and following the development of commercially viable processes for the separation of the rare earths, an intensive research effort, especially in Japan, has led to the filing of a very large number, 700 or more, of patent applications claiming special merit in the use of selected RE metals. As there are 10 of them, the permutations and combinations are vast and it is almost certain that the end of the line has not yet been reached.

The enormous increase in magnetic strength resulting from the development of the RE metal magnets has been an important factor in the miniaturization of electrical equipment, and has opened up a range of entirely new uses for permanent magnets:

Material	Energy product, BH_{max} (kJ m^{-3})
Lodestone	1
Alnico	10–30
Ferrites	up to 30
RE–cobalt	up to 250

Other PM Materials

Copper and its alloys

Oil retaining bearings account for, by far, the bulk of copper alloy parts, but structural parts in copper, bronze, and brass, including nickel brass (commonly called nickel silver), are all produced from powder.

Brass and nickel silver

Pre-alloyed powders are used universally and the most important feature from the PM point of view is the volatility of zinc at the sintering temperatures. If parts are sintered in the normal way in a flow of reducing gas, loss of zinc from the surface takes place. To avoid this it is advisable to keep a special furnace exclusively for brasses and to maintain in it a partial pressure of zinc vapour, or alternatively to sinter the parts in closed boxes, which procedure serves the same purpose. A dry atmosphere is also advisable otherwise oxidation can occur which spoils the appearance of the surface. Lithium stearate instead of zinc stearate is recommended as lubricant, it is said to give better strength and certainly gives a cleaner surface. Compositions include 90/10, 80/20, and 70/30 brasses with or without an addition of up to 2% of lead; and 62/20/18 copper/zinc/nickel.

Aluminium

Aluminium is not an easy metal for PM because of the inevitable coherent film of oxide that covers the surfaces of the particles. Compacts that are sufficiently strong to handle can, however, be produced because, when the adjacent surfaces of two particles are deformed in close contact with each other, the oxide films are ruptured and the exposed clean surfaces weld together. During sintering, however, the oxide films cannot be reduced by even the best sintering atmosphere, and good sintered strength is not achievable. The problem is circumvented by the use of alloy compositions that allow the formation of a liquid phase which breaks up the oxide skins. The alloying elements commonly used are magnesium, copper, and silicon. A low dew point sintering atmosphere is required to reduce the risk of further oxidation.

Opinions among experts have differed about the economic viability of aluminium alloy parts production, but the fact that a significant volume of parts is produced and sold would seem to settle the argument. There are three features that can be cited in favour of aluminium structural parts: firstly, they are broadly speaking corrosion resistant; secondly, they are non-magnetic; and thirdly, they are light in weight compared with iron- or copper-based materials. The overall weight saving in the context of the total weight of, for example, a motor car is insignificant, but for moving parts, the lower inertia is seen as advantageous. This applies equally to business machines.

Stainless steels

Almost the only reason for using stainless steels, which are very much more expensive than carbon or low alloy steels of equivalent strength, is their good corrosion resistance. In a few cases the austenitic grades are used because they are non-magnetic. The reasons for making stainless steel parts by PM are the same as those cited for PM parts generally, i.e. an overall cost saving is the most common. Both ferritic (12% Cr) and austenitic grades are used, among the latter being 18/12 nickel/chromium and a similar composition containing also about 2% of molybdenum. The presence of molybdenum, well established in wrought austenitic steels, increases the corrosion resistance to liquids containing chlorides, e.g. sea water. As regards corrosion resistance, a word of caution is in order; the resistance is not, in all circumstances, as complete as that of the same composition in wrought form, and this aspect is the subject of several current investigations.

In addition to structural parts, stainless steel powders are used, as mentioned earlier, for the manufacture of filter elements for use in corrosive environments or at elevated temperatures.

10 Wrought PM Products and Other Specialities

So far we have discussed mainly the production of individual components from powder, arguing that the ability to do this is the major justification. However, there is another and growing area of PM, namely the production of wrought products. Reference has been made already to the production of tungsten wire, but that is a special case in so far as it is not possible to make wrought tungsten from ingot. The same applies to tungsten's sister metal molybdenum which is now manufactured on a significantly larger scale than tungsten itself. In the same category are certain other metals that have come into use more recently such as beryllium, chromium, and ruthenium, all of which have one feature in common with tungsten, that is marked plastic anisotropy. This means that until the crystals are all suitably orientated, mechanical working is not possible. The metals mentioned find only limited and very specialized applications, and they will not be considered further.

More relevant to our present purpose are materials that have traditionally been made in wrought form from ingot, but which can either be produced more economically from powder or which have improved properties when so produced. The production of dispersion-strengthened alloys with properties, especially at elevated temperatures, much superior to those of traditional alloys is dealt with later. These new materials are at present under consideration for aerospace applications where the benefit of improved properties is so great that cost is not the major consideration.

Superalloys

These comprise a family of alloys designed specifically for load bearing applications at elevated temperatures, the most publicized being in the hot section of gas turbines, in particular for aero-engines. Superalloys were developed from the well known 80/20 nickel/chromium alloy, commonly referred to as Nichrome, although that is in fact one producer's registered trademark. This alloy has for a long time been used in elevated temperature applications: e.g. for the manufacture of wire mesh belts for sintering, brazing, and heat treatment furnaces; for the heating elements of such furnaces and also in domestic electric fires. With the development of jet engines, the demand for greater strength at progressively higher temperatures has encouraged an intense research effort which has led to vastly improved fuel efficiency, among other benefits. The improved properties have come about mainly by including in the composition small percentages of aluminium and titanium which give the alloys precipitation hardenability via the formation of NiAl and NiTi intermetallic compounds, and also by additions of molybdenum and tungsten and the replacement of some of the nickel by its sister element cobalt. Typical compositions are given in table 4.

Table 4

Alloy	Cr	Co	Al	Ti	Nb	Mo	W	B	C	Ni
Astroloy	15	17	4	3.5	—	5	—	0.03	0.02	Balance
Rene 95	12.7	8	3.5	2.5	3.5	3.5	3.5	0.01	0.05	Balance

These compositions are expensive because of the high prices of the constituent metals, but additionally, since they are, by design, resistant to deformation even at elevated temperatures, they are expensive to fabricate. An important economic consideration, therefore, is to get as big a yield of finished component as possible from a given weight of starting material. If the starting material be ingot, there are significant losses all along the line—cropping to remove 'pipe', machining to remove surface imperfections, etc. Furthermore, ingots are prone to segregation so that macro-variations in composition can arise. Because the application is so demanding—aircraft safety depends on it—only consistent material is good enough. On all these counts PM shows to advantage; the drawback, as in most of PM, is that the powder production process is more expensive.

Spherical powders are normally used because they have the highest packing density, and two processes are in commercial use for their production: argon atomization and centrifugal atomization by rotating electrode. These processes were discussed in chapter 2.

Two features of gas atomization that are less than ideal should be mentioned. Firstly, there is the danger of pick-up of refractory particles from the melting crucible or from the nozzle through which the melt flows into the atomizing chamber. These become inclusions in the finished material and are potentially sites for fatigue failure. Secondly, argon bubbles can be trapped inside the particles, and because argon is chemically inert and also insoluble in the metal, it remains as small discontinuities in the wrought product. Various ways have been devised to minimize the number and size of non-metallic particles in the molten metal, including finally passing it through a ceramic foam filter made of alumina or zirconia. Additionally, equipment is now available for removing such particles from the powder before consolidation. Figures 30 and 31 show how the measures that have been taken to improve the process have affected the inclusion count of superalloy powders and wrought material made from powder. It can be inferred, and rightly, that cleanness is perceived as being a major requirement for these materials. The risk of entrapped gas is minimized by rejecting powder greater than 150 μm in diameter since it is in the larger particles that the phenomenon of 'hollow' spheres occurs.

Both the above problems are very largely avoided by using centrifugal atomization with a rotating electrode, which may be melted in near-vacuum by an electron beam or by a plasma arc, but care must be taken to ensure that the electrode material is substantially free from inclusions.

After sieving and testing the powders, the normal process is to load the powder into a container in which it is vacuum degassed and sealed. The whole is then HIPped and hot extruded to billets which may then be further shaped by traditional forging processes.

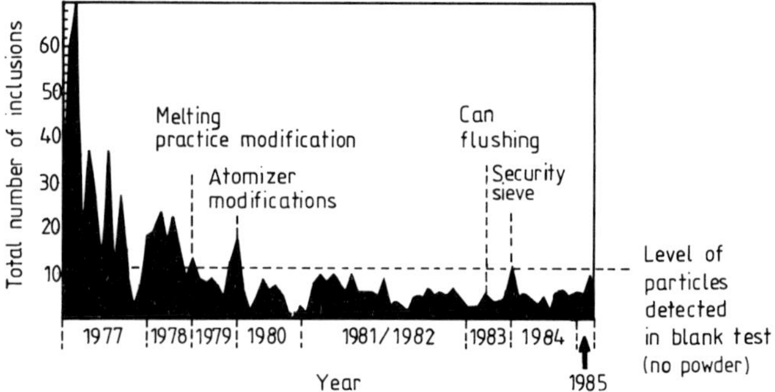

Figure 30 Water elutriation data—NIMONIC alloy AP1 (total number of inclusions in the +106/150 μm fraction of a 0.5 kg sample of blend powder).

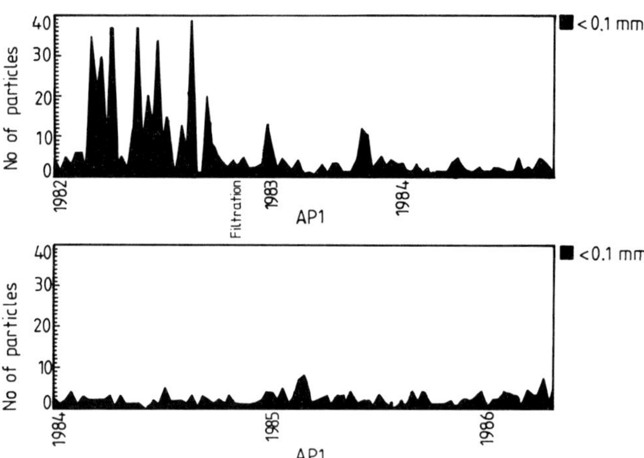

Figure 31 Non-metallic inclusions in extruded powder (size = length after extrusion).

The advantages of the PM route

The advantages claimed for material made by PM include the following:

(1) absence of macro-segregation;

(2) easier working;
(3) better control of grain size, a very important consideration;
(4) more uniform micro-structure and mechanical properties.

Additionally, PM makes it possible to produce new and more complex alloys more easily. The production of so-called near-net shapes in superalloys direct from powder is dealt with later.

High Speed Steels

These are, like superalloys, multi-constituent compositions and are made economically by a process similar to that described for superalloys. Economy results from the much better yield of finished bar compared with the ingot route, and there is also the benefit of improved uniformity of micro-structure, notably the complete absence of carbide stringers in the longitudinal direction (see figure 32). There is also benefit that even more highly alloyed compositions containing, for example, additional vanadium carbide, can be produced.

Other processes for these steels starting with powder are as follows.

(i) Cold iso-static pressing of canned-water-atomized powder, followed by hot extrusion.

(ii) Direct production of individual pieces from irregular powder by either die compaction or CIP followed by carefully controlled liquid phase sintering in vacuum. This process is especially flexible as regards composition since no mechanical deformation is involved at any stage—see example 35, the high speed steel threading die.

Stainless Steels

In the case of these materials the justification for using PM is entirely a question of cost. The process used for rod and especially for

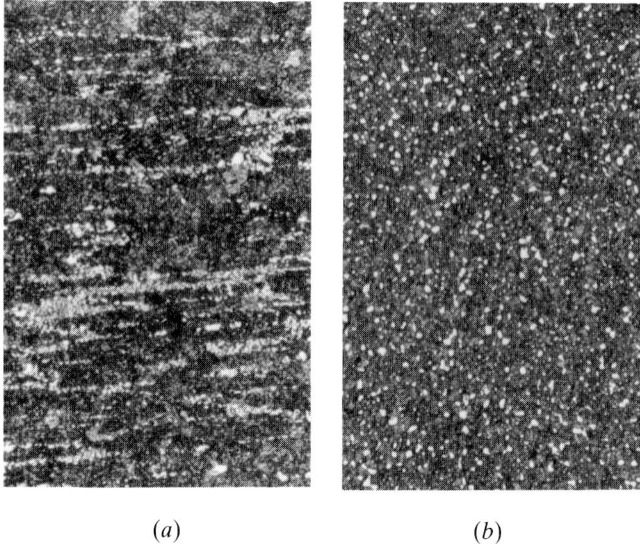

(*a*) (*b*)

Figure 32 Micro-structures of longitudinal section of high speed steel bar. (*a*) Ingot material, (*b*) PM material. *Powdrex Ltd.*

tube is once again to start with spherical atomized powder which is canned and either hot extruded or hot rolled, the yield of finished product more than offsetting the extra cost involved in powder production. For strip a quite different approach has been used—a process called *powder rolling*.

Water-atomized powder is continuously compacted without lubricant by a more or less standard rolling mill. Elaborate precautions are necessary to ensure uniformity of density and thickness across the width of the strip, and also in the subsequent handling of the 'green' strip which is, of course, fragile. The British Steel Corporation has developed and patented a process in which the green strip is carried to and through the furnace on a cushion of gas so as to avoid scratching the surface on the furnace hearth. Argon is the gas used, and it serves also to protect the steel from oxidation. After sintering, the strip is cold rolled to complete the densification and to give it the required high finish. The cost saving in this process comes from the avoidance of the repeated rolling, annealing, pickling, and cropping that are necessary when starting with an ingot. The operations themselves are costly and, additionally, material is lost at each stage. With powder rolling, a very high yield of finished strip is possible.

Powder rolling is also used on a substantial scale in Canada to produce nickel strip for coinage. This is a special case in so far as powder is the end product of the refining process used, so that no powder production cost need enter the equation.

Conform Process

This is yet another process for producing continuous lengths of dense metal from particulate feedstock. The word particulate instead of powder is used intentionally because the process can operate successfully with many different types of raw material, including, for example, chopped swarf and the like. The particulate material is fed by means of a pair of mating grooved rolls, the grooves being provided with contoured surfaces that grip the feedstock which is forced under the pressure generated into a chamber, from which it is extruded through a suitable die. During the process the deformation is such that a significant amount of heat is produced and so, basically, it is warm extrusion that occurs. All types of section that are possible by extrusion can be produced: rods of many sections, wire and strip. Devised initially for the consolidation of scrap material, the process is now being used for the processing of powders made specifically for the purpose. Since the amount of deformation is very large, the process works perfectly well with metals such as aluminium, which have a continuous oxide film on the surface of the particles. During the compression and extrusion the oxide films are broken up, exposing metallurgically clean metal surfaces which readily weld together.

ODS Materials

ODS stands for oxide dispersion strengthening, which is largely self-explanatory. The incorporation of thoria in tungsten and platinum has been referred to in chapter 1, but there the purpose was to enable a particular micro-structure to be developed, not to improve the strength, although it may do that in addition. It is well known that a fine dispersion of a second phase can provide a striking increase in the strength of many metals. This is the basis of the process known as precipitation hardening or age hardening, in which an alloy of two metals which have limited solubility at room temperature is heated to a temperature such that the alloying constituent dissolves in the lattice of the parent metal which is then quenched, retaining the alloying element in meta-stable solid solution. On re-heating the metal to some temperature lower than that for re-solution of the alloying element, the latter is thrown out of solution as a fine precipitate within the grains of the parent metal, thus 'locking' the dislocations on which plastic deformation depends. In the case of certain aluminium alloys, in which the phenomenon was first discovered, the precipitation takes place at ambient temperature if sufficient time is allowed: hence the original name 'age hardening'. A feature of these precipitation hardened alloys is that if they are exposed to elevated temperatures, they gradually soften by over-aging, i.e. the fine precipitate of the second phase coalesces and becomes progressively less effective; indeed it may re-dissolve if the temperature gets sufficiently high. Thus there is a temperature limitation above which such materials cannot usefully operate. In the case of the superalloys used in high temperature gas turbines, the temperature limitation is a major factor limiting the performance of e.g. jet engines. With hindsight, it seems an obvious step to replace the precipitate by a completely insoluble second phase. Like many excellent inventions this one is obvious—afterwards. The key factor is to find a method of introducing the second phase in the necessary very finely divided state. The pioneer work was done many years ago with aluminium, and a product called SAP (sintered aluminium powder) was developed in Switzerland. As we have seen aluminium powder particles inevitably have oxide films on the surface and it was this oxide that was used to provide the improved strength, a powder billet being extensively hot worked to break up the oxide films and disperse them throughout the metal. In order to provide sufficient oxide, flake powder having a large surface area per unit volume was used. Technically the process worked, but commercial exploitation was limited and SAP was eventually superseded by other materials.

However, oxide dispersion strengthening did not die with SAP, and there is currently great interest in ODS superalloys and aluminium. The problem of how to disperse the strengthening particles in the required very finely divided state is ingeniously solved by a process called *mechanical alloying*. This involves ball milling the ingredients—metal powder plus the refractory oxide—in a high energy attritor mill for long periods, during which the oxide particles are hammered into the metal particles, which are alternately welded together and broken up so that the oxide particles are progressively reduced in size. The resulting powder is then processed by canning and extrusion, as already described. Thoria has been replaced by yttria, as the former shows a certain amount of radioactivity which has become unfashionable of late.

Other ODS materials now in commerce are lead and copper. The former is usually strengthened with its own oxide and processed in a generally similar way to that described for SAP. With copper, however, the metal's own oxide cannot be used because it is much too soluble in the metal. Aluminium oxide is used and the necessary finely divided state is achieved by a process known as internal oxidation. An alloy of copper with a small amount of aluminium is atomized and the powder mixed with copper oxide and heated. The oxygen diffuses into the alloy particles and reacts with the aluminium, forming a fine dispersion of Al_2O_3. The powder is then consolidated and extruded. The merit in this case is the provision of increased strength, while retaining substantially the electrical conductivity of pure copper (see figure 33). Incidentally, internal oxidation of an alloy powder is used also to produce the silver/tin oxide and silver/cadmium oxide electrical contact materials mentioned in chapter 1.

Other dispersion strengthened materials

Although oxide dispersion strengthening has occupied the limelight, there is no special magic in oxides. The requirement for good dispersion strengthening is a second phase that can be introduced in the form of very fine particles in the lattice, and which is virtually insoluble in the metal even at the highest temperatures to which it may be exposed in service. A material recently developed in Germany is aluminium with a fine dispersion of aluminium carbide. This is made by mechanical alloying of aluminium powder and graphite. Unlike the mechanical alloying of oxide into superalloys,

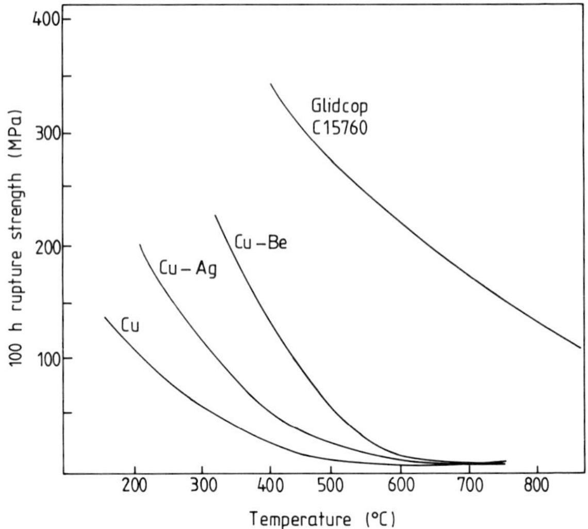

Figure 33 Elevated temperature strength of dispersion strengthened copper compared with other high conductivity copper alloys. *Courtesy SCM Metal Products whose trademark is Glidcop.*

this process could even more accurately be described as alloying in so far as the graphite is, during the milling process, converted to aluminium carbide Al_4C_3. When the alloying is complete the powder is consolidated by CIP and hot extruded. There is inevitably some oxide also in the metal, and a typical composition with the trade name Dispal 2 contains some 14% by volume of total dispersoid. The size of the particles is stated to be between 18 and 85 nm. It is the fineness of the dispersoid that is the key factor as can be seen from figure 34, which shows tensile values for mechanically alloyed aluminium in comparison with SAP. With 14% of dispersoid the room temperature strength is more than three times that of aluminium, but more important is the increased strength at elevated temperatures; e.g. figure 35 shows the results for Dispal 3 compared with standard strong aluminium alloys. Suggested uses for the new materials include pistons and cylinder heads for IC engines, as well as load bearing structural parts for aircraft.

Similarly, improved properties for mechanically alloyed (MA)

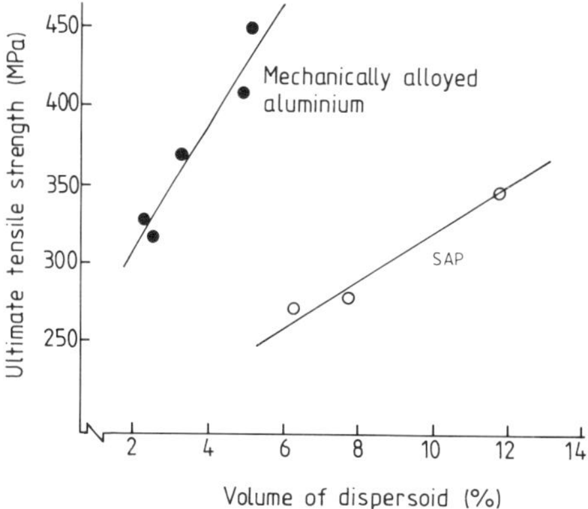

Figure 34 Showing the effect of dispersoid particle size and volume on strength. *Inco.*

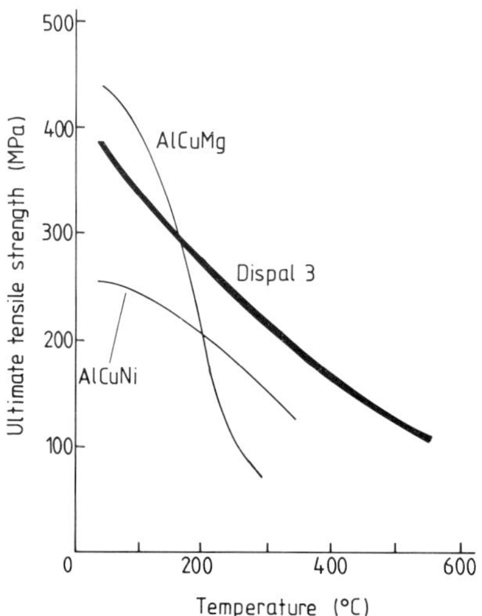

Figure 35 Showing the superior retention of strength at elevated temperature of dispersion strengthened aluminium. *Sintermetallwerk Krebsöge.*

aluminium alloys—Al/4%Mg and Al/4%Mg/1.5%Li—have been reported by Inco.

Production of Near-net Shapes

The processes for the production of PM materials with improved properties, dealt with above, yield for the most part billets which are then brought to the required shape by traditional forging and machining processes. It is logical to ask whether, having powder as the starting material anyway, it is possible to produce the finished article in the same sort of way as with what we may now refer to as traditional sintered parts, i.e. compact and sinter the powder direct to finished size and shape. The potential reward, especially in the aerospace industry, is so great that a very large research and development effort has been made over the last decade. Several processes have been reported for the production of what are called near-net shapes, by which is meant shapes that are sufficiently near to that of the final component that the amount of machining is greatly reduced—as is the amount of scrap produced. Many of these processes are based on pseudo-iso-static pressing at elevated temperature. In one case the powder is loaded into a suitably shaped cavity in a block of metal which, at the pressing temperature, is soft enough to behave almost like a fluid when subjected to uniaxial pressure in a standard hydraulic press. This process has been called *fluid die compaction*. The shaped part is released from the die by melting the latter which can then be re-used. Copper/nickel alloys have been favoured because a wide range of melting points can be made available by varying the composition. A similar pseudo-iso-static effect is claimed in the *Ceracon process* where the pressure is applied via a bulk of special ceramic particles.

Rapidly Solidified Powders

This is another very promising development arising from the discovery that if certain alloys are cooled from the molten state at an extremely high rate, the resulting solid exhibits quite unusual and potentially useful properties, including greatly increased strength. These properties result from a metastable micro-structure which may be micro-crystalline or even amorphous. Cooling rates of the order of a million degrees a second are alleged in some cases, and it will be obvious that rates anywhere nearly approaching that figure can only be achieved in very thin sections. In some processes molten metal is projected against the rim of a rotating wheel made of a metal of good heat conductivity, e.g. copper, which is internally water cooled so that the metal freezes in the form of a very thin ribbon, typically about 20 μm in thickness. Aluminium alloys have been produced by gas atomization using helium, which apart from being chemically inert and, therefore, protective, has by far the best heat conductivity of any possible atomizing gas.

The only way in which these rapidly solidified metals can be used in the manufacture of engineering components is via a PM-type process, i.e. the ribbon material is crushed to powder. In further processing of the powders, care must be taken not to heat them to a temperature at which recrystallization occurs, otherwise the special properties are lost.

Nearest to commercial realization would seem to be rapidly solidified (RS) aluminium alloys, but superalloys are also being extensively studied. Figure 36 shows some of the interesting results that have been reported.

Another metal of great interest to aircraft engineers is titanium. This element is nearer to aluminium than to steel as regards specific gravity, but has a much higher melting point than aluminium and, therefore, retains useful mechanical properties to significantly higher temperatures. The drive to use PM techniques for titanium was, in the first place, to reduce costs by reducing the amount of scrap produced; scrap which cannot readily be re-used. The bulk of the development work was done on the best known standard alloy, Ti–6Al–4V. Both blended elemental powders and atomized pre-alloyed powders were used, and it was found that if the cooling rate of the latter exceeds $10^3\,°C\,s^{-1}$, the same sort of metastable structures and superior properties already demonstrated for RS aluminium alloys could be obtained. This finding led to the investigation of novel alloy systems using alloying elements not normally soluble in titanium, such as the rare-earth metals yttrium, erbium, and neodymium, and very promising results have been reported (see figure 37). Another line of approach that is being pursued

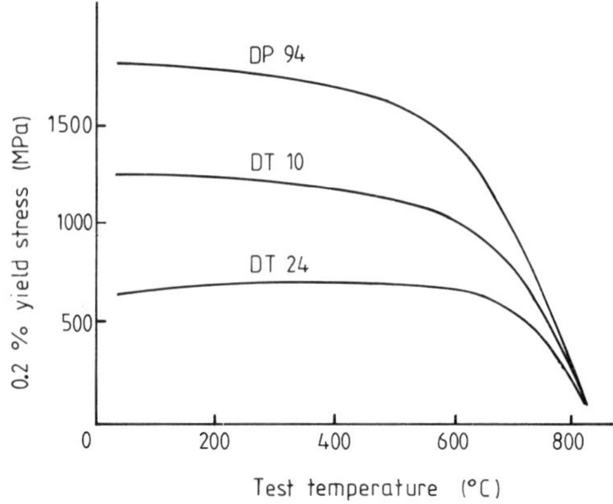

Figure 36 Tensile properties of rapidly solidified Ni-base alloys as influenced by the structure. *ONERA*.

Alloy	Nominal compositions (wt %) Ni	Co	Cr	Ta	Al	B	Structure
DT 24	balance	9	12.2	3.1	2.8	1	Dendritic
DT 10	balance	9.5	14.5	3.3	3.3	2	Micro-crystalline
DP 94	balance	10.4	14.7	3.6	3.6	3.5	Amorphous

is the addition of light metals such as lithium, which is established as an alloying element in aluminium and which gives a worthwhile reduction in density.

Spray Forming

This process, which has made considerable commercial progress recently, is not powder metallurgy in the strict sense of the term in so far as the metal is at no stage in the form of powder. However, for reasons that will be evident, the PM world has adopted it.

The process involves the gas atomization of liquid metal, but instead of allowing the droplets to solidify as powder, the spray is caused to impinge on a solid surface where the liquid or semi-solid droplets are flattened by the impact and solidify, enabling a layer of dense metal to form. This layer may be built up to any desired thickness and, by suitable choice of the design of the target, the

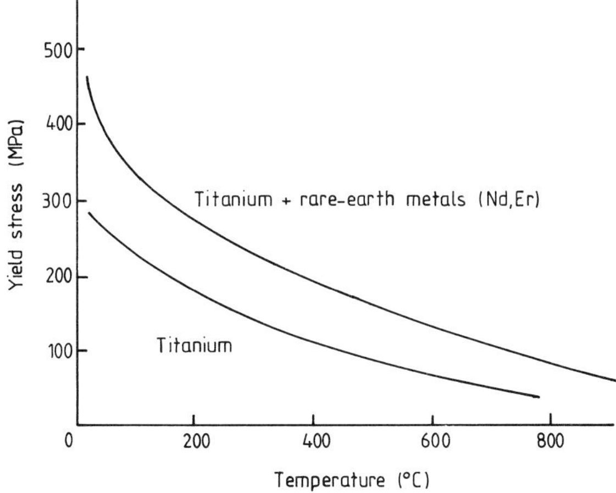

Figure 37 Tensile properties of titanium/rare-earth metal alloys. *US Air Force.*

angle of the spray and other parameters, near-net shapes can be produced. For example, if the target is a cylinder rotating horizontally and capable of being moved in a controlled manner in the axial direction, a tube of deposited metal can be formed. The deposit has all the advantages of dense metal produced from powder, i.e. complete lack of macro-segregation and pipe. Furthermore, by injecting fine refractory powder particles, e.g. by entraining them in the atomizing gas stream, dispersion strengthened material can be deposited.

Examples

Example 1: Shock Absorber Piston

Shock absorber pistons are made by the million and provide a bread and butter product range for the PM industry. No other manufacturing route can compete economically.

The example illustrated is typical. It is 26 mm in diameter and made of plain carbon steel (0.5% C). The final density is 6.3 g cm^{-3}.

The only post-sintering operations are:

sizing
steam treatment
lapping.

Tolerance on the flatness of the surfaces is 0.005 mm.

Example 1 *Sintermetal SA, Spain*

Example 2: Rod Guide for Shock Absorber

This part, weighing 40 g, is made from a 2% Cu/iron alloy, using mixed elemental powders, and has a final density of 6.0 g cm^{-3}. The largest diameter is 25.4 mm.

After sintering, the parts are sized and barrelled.

Mechanical properties:

UTS	160 N mm^{-2}
hardness	HV5 55.
Tolerance	0.02/0.1 mm.

This part replaces an assembly of two separate components, one of which was of plastic.

Example 2 *Manganese Bronze Limited, UK*

Example 3: Belt Pulley

Toothed components such as gears and the like are natural candidates for manufacture by PM since the cost of machining, whether from solid or from a casting, is a very substantial item. With PM, the teeth can be produced very accurately and normally require no subsequent machining.

The part shown is indexed on the shaft of a Renault engine and weighs 420 g. It is made from a 2.5% Cu carbon steel to a final density of 6.5 g cm^{-3} and, apart from sizing, i.e. trueing up the dimensions, no post-sintering operation is involved. The accuracy achieved is IT8.

Mechanical properties:

UTS	280 N mm^{-2}
hardness	95 HV10.

Example 3 *Alliages Frittes—Metafram, France*

Example 4: Gear for a Hand-held Grinder

This quite complicated part was originally made by machining and it will be obvious that the process would be expensive. Production by PM is relatively straightforward and the sintered version costs one quarter of the price of the machined part that it replaced.

The gear is made from a standard diffusion bonded pre-mix containing 4% Ni, 1.5% Cu and 0.5% Mo—plus carbon—and, after sintering, is carbo-nitrided and heat treated.

Diameter	50 mm.
Weight	91 g.
Hardness—core	90 Brinell
—case	700 ± 50 HV1.

Example 4 *Robert Bosch, West Germany*

Example 5: Small Helical Gears

Helical gears are one of the few exceptions to the rule that PM components must have longitudinal surfaces parallel to the pressing direction in order to allow the compacts to be ejected from the die. In the case of helical gears, the punches are caused to rotate during compaction and the punches and the compact rotate during ejection.

The example shown is 21 mm in diameter and 4 mm thick. The angle of the helix is 13°20′.

A partially pre-alloyed, i.e. diffusion bonded alloy steel powder containing 4% Ni, 1.5% Cu and 1.5% Mo is used, enabling a high density to be achieved in a single press and sinter operation. After heat treatment, the part has a hardness in excess of 300 HV1. It is used in the window opening mechanism of motor cars and replaces a moulded plastic part which was in the original design but which proved to have inadequate mechanical properties.

Example 5 *GKN Bound Brook Italia, Italy*

Example 6: Latch Lever for Security Lock

This part—shown on the right—was entirely redesigned for PM to replace a composite component made of steel stampings and plastic—shown on the left. The part is made of copper steel to a density in excess of 7.0 g cm^{-3} and subsequently case hardened to provide much improved wear resistance and, therefore, longer life.

Composition	iron–1% Cu.
UTS	200 N mm^{-2}.
Overall length	71 mm.
Weight	83 g.

The tolerance level of the large hole is IT9.

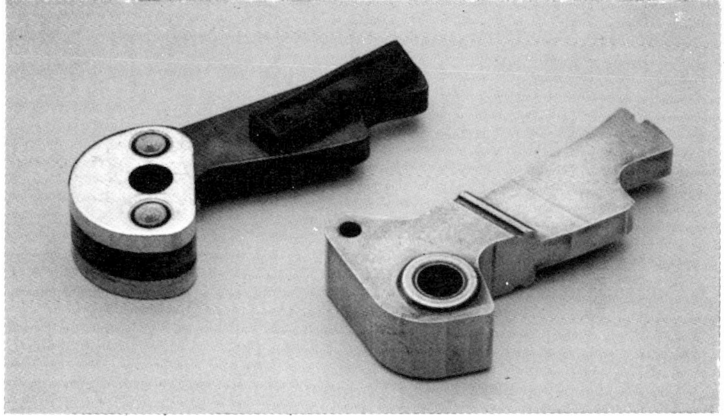

Example 6 *Ringsdorffwerke, West Germany*

Example 7: Coupling Bar for the Injector Pump of a Diesel Engine

This steel part replaces a part made from rolled bar stock by machining. Ten different machining steps were required, including broaching of the main bore and keyway. The design was modified to allow easy production by PM and the only machining required is the drilling and threading of the transverse hole.

The weight of powder used in the manufacture of this part is 140 g, compared with 450 g of steel bar which was required to produce the original machined part.

After sintering, the part is sized and heat treated.

A cost saving of 50% is obtained.

Example 7 *Polmetasa, Spain*

Example 8: Drive Gear for a Chain Saw

The composition of this part is iron with 1% Cr, 0.8% Mn, 0.3% Mo and 0.6% C, which composition is heat treatable and gives the good mechanical properties required, namely:

UTS	550 N mm^{-2}, as sintered
hardness after HT	900 HV3
OD	36 mm
weight	33 g.

This part has replaced the cast steel part formerly used and provides better wear resistance, as well as a cost saving.

Example 8 *Ringsdorffwerke, West Germany*

Example 9: Support for Transmission Control Lever in Passenger Cars

This complex component is an assembly of two sintered parts and involves an unusually large number of operations. Nevertheless, compared with a traditionally forged and machined part of almost identical geometry, a cost saving of 43% was reported. This saving results from the markedly reduced amount of machining required on the PM part. The PM assembly and the traditionally forged part were used interchangeably over several years to assess performance before the user switched fully to the PM version.

The material is a 2% Ni steel containing 0.6% MnS which confers free machining qualities that were found to be necessary to reduce tool wear, especially of the expensive broaches used to produce the splines in the main bore.

The powder mix is compacted to a density between 6.85 and 7.05 $g\,cm^{-3}$, and the compacts are sintered at the relatively high temperature of 1280 °C, in order to get good diffusion of the nickel.

Example 9 *Sintermetallwerk Krebsöge, West Germany*

The two parts—shown on the left—are then projection welded together. The user's requirement was for the weld to withstand a shear load of 10 kN without plastic deformation or fracture. In practice, fracture loads of 40–50 kN are achieved in the heat treated component.

The complete manufacturing sequence is given below.

The complete process

(1) Compact the thinner section.
(2) Sinter.
(3) De-burr by shot blasting.
(4) Compact the main section.
(5) Sinter.
(6) De-burr by shot blasting.
(7) Turn the radial groove, the lateral flat, and two height tolerances on main section—in one operation.
(8) Projection weld the two sub-components.
(9) Drill two radial bores, one being countersunk.
(10) Countersink both ends of the central bore.
(11) Broach two splines and the diameter of the central bore.
(12) Wash in trichloroethylene to remove broaching oil.
(13) Carbo-nitride and temper.
(14) Wash in trichloroethylene to remove quenching oil and soot, if any.
(15) Hone the central bore 0 15 H7.
(16) Hone the lateral bore 0 6 H7.
(17) Wash to remove honing oil.
(18) Press fit hardened and ground pin into lateral bore, and two PTFE coated bushes into the central bore. (These are the parts shown on the bottom right of the picture.)
(19) Inspect.
(20) Apply a corrosion protection dip.

Example 10: Synchronizer Ring for a Truck

This comparatively large part is a sinter forging. It replaces a conventionally forged steel part of similar design, and affords a cost saving estimated at some 50%. This saving results from the superior accuracy of the sinter-forged part which needs no machining other than grinding of the cone. The part is subsequently case hardened.

Composition	Iron with 2% Ni, 0.25% Mn, 0.5% Mo, 0.25% C.
Density	7.7 g cm^{-3}
Weight	830 g.
OD	800 mm.
Hardness	180 Brinell
UTS	800 N mm^{-2}.
Fatigue strength	300 N mm^{-2}.
Elongation	10%.

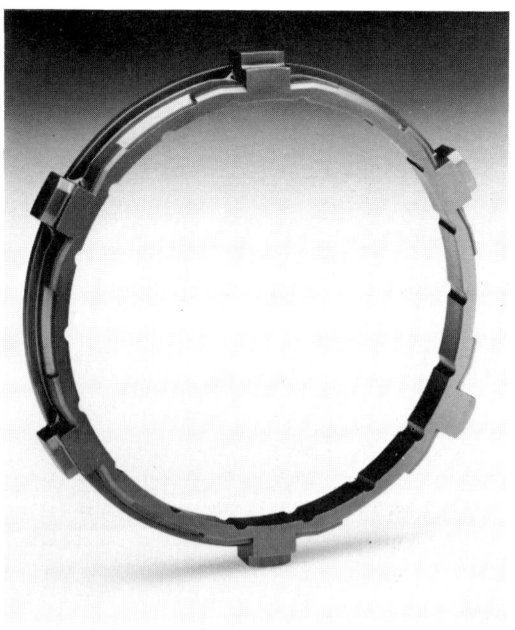

Example 10 *Sintermetallwerk Krebsöge, West Germany*

Example 11: Soft Magnetic Stator for a High Speed Printer

This part is made of a 3% Si/iron alloy to a density of 7.4 g cm^{-3}.

After sintering, three faces are ground to fine tolerances, four holes are drilled and one of them threaded. The part is then zinc plated.

Dimensions	40 × 38 mm and 7 mm thick.
Weight	34 g.
Strength	400 N mm^{-2}.
Hardness	75 RB.

The sintered part is easier to machine than a similar part cut from plate, and a cost saving of some 10% results.

Example 11 *AMES, Spain*

Example 12: Rear View Mirror Holder

This consists of two parts fitted together after sintering and sizing. It is glued to the windscreen of motor cars and is made from grade 316 stainless steel. This composition was chosen not primarily for its corrosion resistance, but because it has a coefficient of thermal expansion much nearer to that of glass than that of alternatives such as die-cast zinc or aluminium, and it has the required strength.

Composition	18% Cr, 12% Ni, 2% Mo, balance iron.
Weight	22.6 g.
OD of larger part	32 mm.
ID of smaller part	9 mm to tolerance IT7.
Density	>7.0 g cm^{-3}.
UTS	>400 N mm^{-2}.

The picture shows the two separate parts on the right, and the final assembly on the left.

Example 12 *Ringsdorffwerke, West Germany*

Example 13: Synchro-hub

This toothed component is for the manual gearbox of a motor car. It is made in an alloy steel containing 5% Ni, 2% Cu and 0.5% Mo, with a carbon content of 0.5%. The weight is 245 g.

Sintered to a final density of $7.0\,g\,cm^{-3}$ the following mechanical properties are reported:

yield strength	$560\,N\,mm^{-2}$
UTS	$800\,N\,mm^{-2}$
elongation	2%
hardness (Vickers)	450.

Dimensional accuracy IT8.

The friction surfaces are machined to a roughness value of 1.2 μm.

Example 13 *Alliages Frittes—Metafram, France*

Example 14: Brass Dead Bolt

Replacing a forging that required considerable machining and still had assembly problems, this sintered brass component is used in a commercial mortice lock. Apart from sizing and press fitting of the pin in the end section, no secondary operation is required.

Tests include an impact test in which the part withstands an impact force of 150 ft lb and, when located in the door strike plate, a shear force of 1200 lb.

Critical tolerance is on parallelism, and is better than 0.010″ (0.254 mm).

Example 14 *Presmet Corporation, USA*

Example 15: Hammer Block for High Speed Printer

This soft magnetic component supports 15 hammers in a large computer, and is made from 'pure' iron. It is an assembly of three sintered parts joined together by projection welding.

Fifty holes are drilled and threaded, three faces ground, and the whole then steam treated. Finally the two pins are press fitted into the assembly.

Dimensions	$114 \times 56 \times 14$ mm.
Weight	443 g.
Density	6.8–7.0 g cm^{-3}.
Strength	210 N mm^{-2}.
Hardness	15 RB.

The alternative production process is precision casting followed by extensive machining, but the cost of each component produced in this way was 15 US dollars more than the PM component.

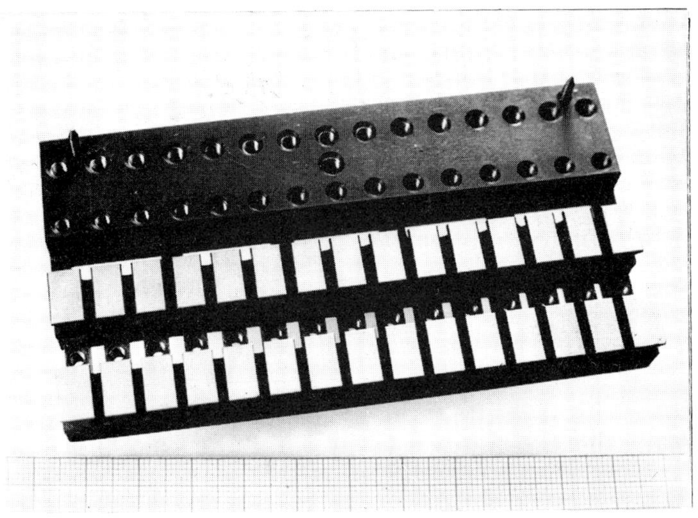

Example 15 *AMES, Spain*

Example 16: Automotive Pump Gears

Both these parts are made from a carbon-containing Cu, Ni, Mo steel, and are heat treated.

The inner gear weighs 140 g and has a density of $\nless 7.0\,\mathrm{g\,cm^{-3}}$. The UTS is $\nless 600\,\mathrm{N\,mm^{-2}}$, and the hardness is >220 HV10.

The outer part, which weighs 180 g, has a density $\nless 6.6\,\mathrm{g\,cm^{-3}}$, a UTS $\nless 520\,\mathrm{N\,mm^{-2}}$, and a hardness >180 HV10.

The only finishing operation needed after hardening and tempering is grinding of the flat surfaces.

Example 16 *FIPS SpA, Italy*

Example 17: Gear Assembly

This assembly of PM steel parts is used in the gearbox of a high performance motor car.

Example 17 *Hitachi Metals, Japan*

Example 18: Platen Adjusting Handle

This part, used to adjust the high speed printer platen of an office computer, is made in a nickel steel. Of especial interest are the several different levels, the production of which requires very careful tooling.

The part is heat treated to a hardness of 35–42 RB and given a black oxide coating.

The tolerance on the ID of the hub is ±0.05 mm.

The PM component replaces a similar part made as an investment casting and results in a cost saving of 40%.

Example 18 *Pacific Sintered Metals, USA*

Example 19: Connecting Rod and Bush for a Refrigerator Compressor

These parts, made of 2% Cu iron, replace similar parts made from cast and machined aluminium. Apart from an economic advantage, the sintered components have better strength and have allowed the compressor to be up-rated to provide a higher output.

The length of the rod is 47 mm and close tolerances are called for. The specified hardness is 75–90 HRB.

After sintering and sizing, the parts are steam treated, primarily to increase the wear resistance but also to confer some degree of corrosion resistance.

Example 19 *GKN Bound Brook Italia, Italy*

Example 20: Multi-level Clutch Hub

In the traditional manufacture of sintered clutch hubs for motor cars with manual gear change it is common practice to finish the grooves on the outer rim by machining. The pressure to reduce size and weight led to a requirement for even higher strength, and this was found to make such machining more difficult and expensive.

Special tooling enabled the production of the part illustrated to keep the amount of machining to a minimum, and consequently the cost also is significantly reduced.

Example 20 *Sumitomo Electrical Industries, Japan*

Example 21: Claw Clutch

This part is used in the drive of agricultural machines.

The material is a 1.75% Ni, 1.5% Cu, 0.5% Mo steel, and when compacted to a green density of $6.8\,g\,cm^{-3}$, the balance of the alloying elements enables the density and, therefore, the dimensions to remain unaltered after sintering. Dimensional accuracy is to IT10.

The part is carburized and heat treated giving the following core properties:

UTS	680–780 $N\,mm^{-2}$
yield strength	580–650 $N\,mm^{-2}$.

Example 21 *SAMO, Italy*

Example 22: Sear Housing in a Pistol

This nickel steel part is made by injection moulding, a technique particularly appropriate for small components of complex geometry.

After sintering, the following mechanical properties are reported:

yield strength	150 000 psi (1034 N mm^{-2})
UTS	200 000 psi (1379 N mm^{-2})
hardness	56 RC.

The part was previously machined from solid, and the elimination of the machining steps led to a cost saving of more than 60%.

Example 22 *Multimaterial Molding, USA*

Example 23: Starter Pinion for a Lawnmower

This part was designed specifically for manufacture by PM.

The material is an alloy steel containing 0.4% C, 2% Ni, 2% Cu and 1% Mo. It is compacted to a green density of 6.7 g cm^{-3}, which increases to 6.8 g cm^{-3} after sintering. The UTS is 470 N mm^{-2}, and the Vickers hardness is 290.

Finally, the teeth are induction hardened.

Example 23 *Alliages Frittes—Metafram, France*

Example 24: Clamp Actuator

The picture shows four views of this part which is used in the clamping mechanism of the engine shroud of a fighter aircraft.

It is made in a 17-4PH stainless steel, and provides a minimum yield strength of 130 000 psi (896 N mm^{-2}) and UTS of 150 000 psi (1034 N mm^{-2}).

The original design was for an investment casting, but by using PM and thereby eliminating the need to machine the cross holes, a cost saving of 50% was achieved.

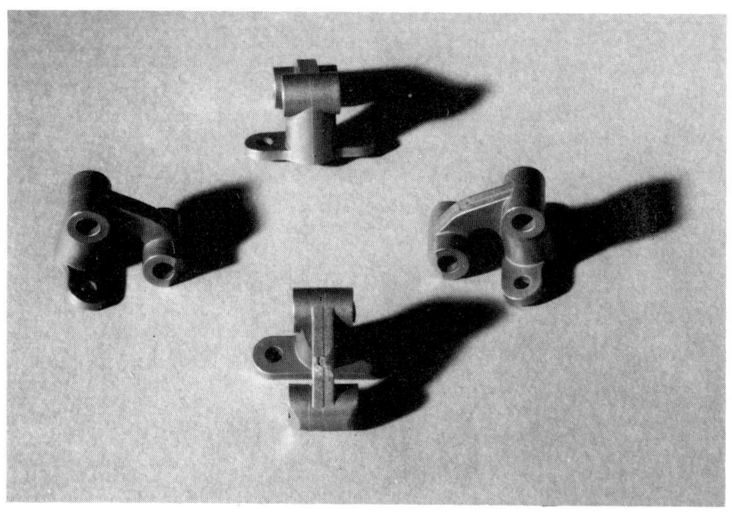

Example 24 *Form Physics Corporation, USA*

Example 25: Cutter Plate for Meat Grinder

This part, made from a Ni, Cu, Mo steel, replaces a part machined from stainless steel plate and affords a cost saving of some 50%. It is pressed and sintered to a final density of $7.1\ g\ cm^{-3}$.

With a carbon content of 0.6% and the relatively high density, this alloy steel has good mechanical properties after heat treatment:

UTS	$1100\ N\ mm^{-2}$
hardness	300 HV10.

After heat treatment the flat faces are ground and the part is then steam treated.

Example 25 *Tozmetal, Turkey*

Example 26: Yoke Coil Support for Dot-printer

Made from carbonyl iron with cobalt and a small percentage of vanadium, this soft magnetic part is sintered to a density >99% theoretical, in order to achieve the high induction value required.

It replaces at a considerable saving in cost an assembly of single yokes made by precision casting.

Example 26 *Sintermetalle Prometheus, West Germany*

Example 27: Projection Welded Lever

The picture shows two versions of a lever for the lifting attachment of a farm tractor. The one on the left is the PM version while on the right is the steel forging that it replaced with a cost saving of 50%.

The especial interest of this component is the shape. It will be obvious that the whole could not be compacted in nor ejected from a single die, and the solution was to make two separate parts and join them by projection welding. The projections are moulded on the surface of one of the parts, and a special jig is used to ensure correct alignment.

The material is a Cu, Ni, Mo steel and the final density is $6.8\,g\,cm^{-3}$. The finished assembly is carburized and heat treated. This gives a UTS greater than $650\,N\,mm^{-2}$ and a case hardness of 64 HRA.

Example 27 *Tecsinter, Italy*

Example 28: Soft Iron Pole Piece

The material used for this part is 0.45% P iron. Of particular interest is the square section, the accuracy of which is better than that achieved in parts made from solid steel. PM is also the more cost effective.

Maximum dimension	30 mm—tolerance 0.1 mm.
Weight	48.6 g.
Density	6.4 g cm^{-3}
Strength	230 N mm^{-2}

Post-sintering operations are limited to sizing and barrelling.

Example 28 *Manganese Bronze Limited, UK*

Example 29: Swash Plate

This component acts as a crankshaft in the compressor of the air conditioning unit of a Ford motor car. Special tooling was successfully designed to cope with the 23° angle between the plate and the hub, and the part was made to finished size, requiring no subsequent machining. It replaces a part made of ductile cast iron which needed four separate machining operations.

Material	alloy steel with 0.5% Ni, 0.6% Mo, 0.3% Mn, 0.5% C.
Density	6.4 g cm^{-3}
UTS after induction heat treatment	565 N mm^{-2}
Tolerances	—on hub ID ±0.002″ —on overall height ±0.004″.

Example 29 *Ferralloy, USA*

Example 30: Indexable Cutting Tool Inserts

The bulk of cutting tool inserts is made from hardmetal (cemented carbide), but for some applications high speed steel is a better choice on economical and, in some applications, on technical grounds, e.g. in the milling of tough high strength steels, in the interrupted cutting of austenitic steels and nickel-base alloys.

The parts illustrated are of high speed steel.

High cobalt, high vanadium grades, such as M35, T15 and T42, are usually specified.

The normal manufacturing route is:

compact
sinter to full density in vacuum (liquid phase sintering)
anneal and heat treat
grind.

An alternative that is also employed involves lower temperature sintering to 97–98% density, followed by HIPping to full density. This retains a finer grain structure which some users prefer. Extensive machining trials, however, have not confirmed any advantage in performance for the finer grained material.

Example 30 *Sintermetallwerk Krebsöge, West Germany*

Example 31: Mould for the Top Section of Glass Bottles

The mould required to produce the lipped or threaded top section of glass bottles needs to be in two halves to enable the bottle to be extracted.

The components illustrated were originally made of grey cast iron, and had a life of around 80 000 bottles. Replacing the cast iron by sintered T15 high speed steel vastly improved the life—the mould being still in use when counting was discontinued after half a million bottles had been produced.

The major benefit in this case is, of course, the reduction in down-time of the machine.

Example 31 *Sintermetallwerk Krebsöge, West Germany*

Example 32: Ejector Pads for Ice Maker

This part is used in a refrigerator designed for the automatic production of ice 'cubes'. Straightness and flatness are critically important: the centre line tolerance is only 0.005″ along the whole 5.5″ length.

The material used is austenitic stainless steel (316L). Compacts are sintered in cracked ammonia at 1260 °C and sized to a final density of 6.6 g cm^{-3}. Barrel finishing ensures the smooth surface required to prevent the ice from sticking to the pads. The hardness is 65 RB.

The part replaces one of similar geometry made as a forging from aluminium which was finally anodized. The stainless steel part proved to have a far better fatigue life, operating without failure for 100 000 cycles. The aluminium part had a life between 20 000 and 45 000 cycles. The corrosion resistance of the steel is an additional benefit: the aluminium corroded when the anodized surface layer wore off.

Example 32 *Remington Arms Co., USA*

Example 33: Titanium Compressor Stator Connecting Link

Because of its exceptional strength-to-weight ratio coupled with excellent corrosion resistance, titanium is used to a significant extent in aircraft where weight is a primary consideration. The metal is expensive and the recovery of machining swarf for direct re-use is not feasible, and so the cost of components machined from bar etc is increased proportionately. PM provides a production route that minimizes scrap production.

The part illustrated was made for a Pratt and Witney engine, and the powder was pressed and sintered to finished shape so that very little machining was necessary. The overall cost saving compared with production from wrought bar was 77%.

The mechanical properties were equal to those of the wrought product.

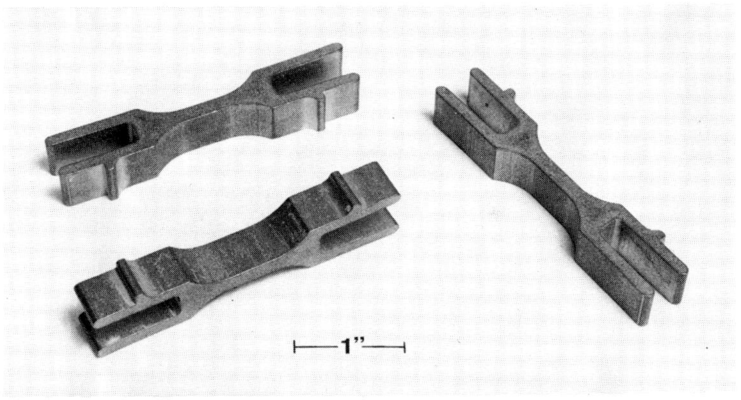

Example 33 *Clevite Industries, USA*

Example 34: Valve Body and Plug

Required to operate in a corrosive environment at elevated temperature, this valve is made of zirconium. The body weighs 4.5 lb and was originally made from bar by machining. 10 lb weight of material was required. By changing to PM the weight of input material was reduced to 6.2 lb—a well worthwhile economy of this very expensive metal.

Pre-forms are made by cold iso-static compaction followed by sintering and HIPping to full density. The longitudinal bore of the body and the bore of the plug are made directly to final size, thus materially reducing the amount of machining necessary.

The mechanical properties and the corrosion resistance of the PM parts are similar to those of the wrought material that they replace.

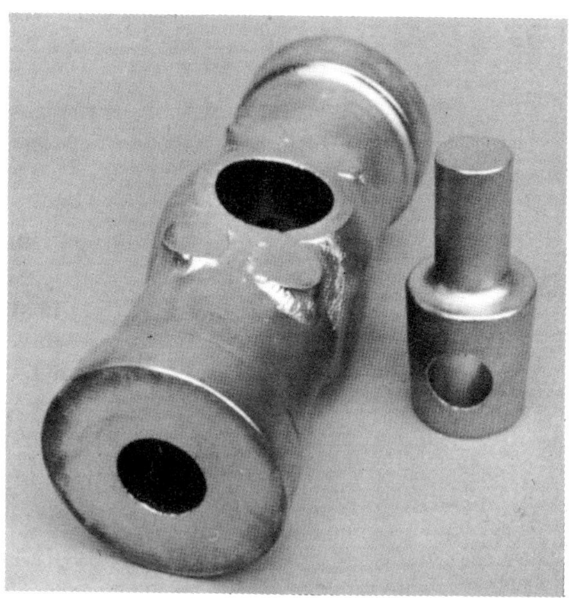

Example 34 *Teledyne Wah Chang, USA*

Example 35: High Speed Steel Threading Die

This part is 25 mm in diameter and weighs about 25 g.

It is produced from water-atomized powder which is cold compacted and liquid phase sintered to full density in vacuum.

There are significant technical advantages in making the part by PM. Conventionally, chip clearance holes are formed by drilling and are, therefore, circular; using powder metallurgy, it is possible to profile the holes to provide the optimum angle for threading. Additionally, with bar material there is a risk of poor quality at the centre due to residual pipe; and it is, of course, the centre that is the most important area. The PM die has uniform cleanliness and properties throughout the section.

Cost data given below show a very large saving for dies made from powder.

Example 35 *Powdrex Limited, UK*

From bar:	
cost of M2 HSS bar (39 g)	19
labour	19.1
other variable costs	25.5
works overheads (200% of labour)	38.2
total per part	101.8.

From powder:	
cost of powder (26 g)	10.5
labour	2.1
other variable costs—based on batches of 5000	6.0
overheads (200% of labour)	4.2
total per part	22.8.

The figures are in pence sterling.

The examples shown in the preceding pages do not reflect the fact that, as regards tonnage, the automotive industry is by far the biggest user of sintered parts. They have been selected to show the wide range of components that are being produced and the many areas in which they find application.

The information given was provided by the parts' manufacturers, to whom the author would like to express his gratitude.

Further Reading[†]

Cemented Carbides K J A Brookes (East Barnet, Herts: International Carbide Data)

Handbook of Powder Metallurgy 2nd edn H H Hausner and M K Mal (New York: Chemical Publishing Corporation)

Liquid Phase Sintering R M German (New York: Plenum)

Powder Metallurgy—Criteria for Design and Inspection Enrico Mosca (Turin: Associazione Industriale Metallurgici Meccanici Affini)

Powder Metallurgy Design Manual 1989 (Princeton, NJ: Metal Powder Industries Federation)

Powder Metallurgy—Principles and Applications F V Lenel (Princeton, NJ: Metal Powder Industries Federation)

Powder Metallurgy Science R M German (Princeton, NJ: Metal Powder Industries Federation)

For keeping up to date:

Metal Powder Report—the monthly journal of MPR Publishing Services

[†] All these publications can be obtained from MPR Publishing Services Limited, Old Bank Buildings, Bellstone, Shrewsbury, Salop SY1 1HU, UK.

Index